Electricity Services in
Remote Rural Communities

Electricity Services in Remote Rural Communities
The Small Enterprise Model

Teodoro Sanchez

ITDG
PUBLISHING

Intermediate Technology Publications Ltd.
Schumacher Centre for Technology and Development
Bourton on Dunsmore, Rugby,
Warwickshire CV23 9QZ, UK
www.itpubs.org.uk

Intermediate Technology Publications Ltd. is the wholly-owned publishing
company of Intermediate Technology Development Group Ltd (working
name Practical Action). Our mission is to build the skills and capacity of
people in developing countries through the dissemination of information in
all forms, enabling them to improve the quality of their lives and that of
future generations.

Typeset in Trade Gothic and Stone Serif by
S.J.I. Services
Printed by Replika Press

Contents

Figures, tables, maps and plates

Foreword

This book appears at a time when community-based rural electrification is the subject of intense activity across many developing countries, particularly in Africa and Asia. Access to electricity is seen as vital for social and economic development. However, in most developing countries there are no sound models for the management of community-based rural electrification schemes, and there is a high risk of failure, even when the large utility companies are involved. Many schemes collapse in conflict and recrimination because they generate too little revenue to cover running costs. Fortunately many of the basic lessons have already been learnt and this book describes the methods that have been developed over the last 20 years by trial and error in Peru, where there have been successful community-based rural electrification projects based on micro-hydro power.

We now understand some of the key factors that are crucial to the success of these projects. The most important point is that the technical installation is just the start. The real challenges are in the detail of how communities, operators and owners can work successfully together without conflict. While every country, indeed every community, is likely to be different there are some basic ground rules that have been proven to work in several different settings. This book provides the framework for the type of detailed agreements that are needed if there is to be an atmosphere of trust that will result in a sustainable system, where bills are paid, a reliable service provided and the costs covered.

This book is highly recommended as a guide for anyone wanting to promote rural electrification in remote communities. The many rural electrification agencies being set up will find it a very practical guide to ensuring that electrification schemes can be managed successfully, whether owned by private operators, municipalities or communities.

Ray Holland
EU Energy Initiative
Partnership Dialogue Facility
Deutsche Gesellschaft für
Technische Zusammenarbeit (GTZ)

Acknowledgements

The model for organizing small-scale rural electricity services, described in this book, was researched, designed and tested through the hard work of the Practical Action energy team in Latin America, which I was privileged to lead for almost 10 years. Many thanks to Luís Rodriguez, Rafael Escobar, Saul Ramirez, Gilberto Villanueva, Janet Velazquez and José Zambrano. The team's work was made possible by the institutional support of the director, Alfonso Carrasco, who not only encouraged the programme's initiatives, but also actively contributed to the development of the book. To all these people, I offer my sincere thanks. I am also grateful to Dr Cecília Flores, who reviewed the legal aspects of the Code of Conduct.

The research into organizing small-scale stand-alone energy schemes in Peru between 1997 and 1999 was funded by the Energy Sector Management Assistance Programme (ESMAP) of the World Bank, whose support was greatly valued.

Finally, many thanks to Dr Nigel Smith and Dr Arthur Williams, my tutors at Nottingham Trent University on my PhD research, from which has been taken the background information for this book.

Teodoro Sánchez
Rugby, 2006

Acronyms and abbreviations

ADINELSA	Empresa Administradora de Infraestructura Eléctrica SA (Company for the Administration of Electrical Infrastructure)
CONSUCODE	Consejo Superior de Contrataciones y Adquisiciones del Estado (Superior Council for State Contracts and Acquisitions)
CTE	Comisión Nacional de Tarifas (National Commission of Tariffs)
ESMAP	Energy Sector Management Assistance Programme
ETW	Energy for Tomorrow's World
GEF	Global Environment Facility
GTZ	Deutsche Gesellschaft für Technische Zusammenarbeit
IDB	Inter-American Development Bank
ITINTEC	Instituto de Investigación Tecnológica Industrial y de Normas Técnicas (Institute of Research on Industrial Technology and Technical Norms)
MEM	Ministério de Energía y Minas (Ministry of Energy and Mines)
OLADE	Organización Latino Americana de Energía (Latin American Organization for Energy Development)
OSINERG	Organismo Supervisor de la Inversión en Energía (The Energy Investment Supervising Bureau)
PESENCA	Programa Especial de Energía de la Costa Atlantica (Special Programme for Coastal Atlantic Energy)
PROMIDEHC	Programa de Desarrollo de Microcentrales Hidroeléctricas (Hydroelectric Microcentral Development Programme)
PRONAMACHCS	Programa Nacional de Manejo de Cuencas Hidrográficas y Conservación de Suelos (National Programme for Managing Hydrographic River Basins)
PROPER	Programme for the Promotion of Renewable Energies
PV	Photovoltaic
RUC	Regimen Unico de Contribuyentes (Unique Registry of Contributors)
UNDP	United Nations Development Programme
VAT	Value Added Tax
WEC	World Energy Council

Introduction

Access to electricity is recognized as an essential component of development but in rural areas there are no simple solutions to providing this service. Extending the national grid is an expensive solution that can be inappropriate in rural areas. Stand-alone schemes present a possible answer, but the high costs and sustainability problems they face have made them unpopular. This publication points out the critical factors for successfully implementing stand-alone energy-generation schemes in remote rural areas. It offers a model for organizing and managing these schemes successfully and sustainably.

The book is organized in two parts. The opening chapter of Part I outlines the international context and the main challenges of rural electrification. It describes the progress of rural electrification in Latin America and in Peru. It also provides a background to how the management model for rural stand-alone electrification schemes was developed. This model is the focus of Chapter 2, describing the work of Practical Action to develop and test it as part of research and development on electricity services in remote rural areas. The model was designed with a clear aim in mind: to promote efficient financial and technical management of electricity services within the social and economic environment, seeking the committed participation of the community. The model incorporates the concept of private management – a micro-enterprise responsible for running a scheme receives payment in exchange for managing the service. It uses a new tariff structure for rural areas (the system of descending blocks) and its application involves the consideration of the model's technical and social viability, as well as the legal aspects affecting the different actors. Examples include municipal laws, environmental regulations, the laws on electricity concessions and national electricity standards.

To date, the management model has been piloted in five small-scale hydroelectric plants in Peru, and in all cases the results have been positive in financial, social and technical terms. The plant in Conchán (40 kW) has worked uninterruptedly with this model since 1999, Tamborapa (40 kW) since 2000, Las Juntas (25 kW) since 2000, Huarango (30 kW) since 2002 and Huarandoza (180 kW) since 2003. In each case, the management model and tariff structure have promoted the use of energy for income-generating and productive activities. Chapter 3 concludes Part 1 with a case study of the pilot project in Conchán, reviewing the lessons to emerge from the scheme. Part II of the book presents the four instruments required to apply the management model. First, the tariff structure, derived from previous studies of socio-economic conditions and the development potential of electricity in the region, is described. Second, the Code of Conduct is presented. This highlights the responsibilities of the different actors and the activities that need to be performed for the scheme to function smoothly. Third, contracts between the owner of the generating scheme and the service operator are given, and, finally, contracts between the service operator and the users are presented. Organization is a key theme for the sustainability of rural electricity services. Therefore, despite the success achieved in the application of this model, it is hoped that, as experiments continue, further modifications and additions to the concepts and instruments will emerge.

PART I
Managing rural electrification

CHAPTER ONE
Rural electrification

Rural electrification worldwide

Access to electricity

It is widely recognized that electricity is a central element for social and economic development. It contributes to economic growth and underpins a range of basic welfare services, such as clean water, health, communications and education. The World Energy Council (WEC) states that 'the message of ETW [Energy for Tomorrow's World] was to remember that health, water, food, education, and many other key aspects of welfare can not be improved unless modern energy becomes available to all' (WEC, 2000:1). Energy plays a key role in providing other basic services in the developing world, and its promotion is therefore a high priority:

> The degree of development of a society can be expressed in terms of increasing ability to meet a certain number of fundamental needs. They can be basic food, housing, clothing, or more elaborate ones like education, culture, exercise of civil rights, quality of the natural environment, leisure, etc. But most of them require energy in varying degrees, thus making the availability of energy an absolute pre-requisite to economic and social development.
>
> (Mariginac and Schneider 2001:81)

Providing access to electricity is one of the most difficult tasks that governments in developing countries confront. Nearly one-third of the world's population had no access to electricity by the end of the 20th century, the great majority of them living in developing countries (Barnes et al, 1997; WEC, 2000). Despite efforts to increase access to electricity services in developing countries and increasing electricity coefficients (percentages of coverage), the total number of rural inhabitants without access to such a basic service remains high. Table 1.1 shows that, by 1990, the number of rural people without electricity was nearly 1.65 billion (67 per cent of the total rural population). Adding the 350 million urban people without access to electricity, this amounts to a total of about 2 billion people without this service.

While the actual number of people without electricity in the cities and urban areas of developing countries remains very high, by far the greatest numbers of people without electricity live in rural areas. Indeed, the challenges of connecting urban populations to existing electricity grids are considerably less

Table 1.1 Percentage of urban and rural population connected to electricity in developing regions

Region	Urban (%)		Rural (%)	
	1970	1990	1970	1990
North Africa and the Middle East	65	81	14	35
Latin America and the Caribbean	67	82	15	40
Sub-Saharan Africa	28	38	4	8
South Asia	39	53	12	25
East Asia and the Pacific	51	82	25	45
All developing countries	52	76	18	33
Total served in developed countries (millions)	320	1100	340	820

Source: World Bank (1996)

than connecting those people who live in remote areas. The uneven growth of electrification in rural and urban areas is a prominent trend. Table 1.1 shows that rural electrification coverage rose from 18 per cent to 33 per cent between 1970 and 1990, whereas for the same period urban electrification appeared to slow down. However, if we count the number of new connections, the figures show the opposite: in urban areas coverage increased from 320 million to 1100 million (a 3.5-fold increase) and in rural areas from 340 million to 820 million connections (a 2.5-fold increase).

Uneven access to electricity within continents and regions (World Bank, 1996; WEC, 2000), within countries and within zones and areas in each country is prominent. For example, Table 1.1 shows that in 1990, whereas on average in sub-Saharan countries electricity coverage was only 8 per cent, in East Asian and Pacific countries it was 45 per cent, and in Latin American and Caribbean countries it was 40 per cent. To decrease the inequality of access to electricity between urban and rural areas, high levels of economic investment are needed, presenting a considerable challenge to governments and development agencies globally:

> The World Bank currently estimates the total energy investment need in Africa, to 2010, at US$110 billion; this is approximately the 1992 cost of German reunification, or the total flow of all aid monies to the Third World in 1990. The World Bank has indicated that it can afford only 10 per cent of investment indicating clearly that private capital will need to finance a significant remainder.

> (O'Keefe, 1996:12)

However, uneven progress in terms of access to electricity is not only attributable to the lack of economic resources in poor countries or to the unwillingness of governments to invest in rural energy; there are other important barriers that prevent the progress of electrification, especially for rural areas (Platow and Goldsmith, 2001). Although authors differ about the number of barriers and their nature, they relate to the social, economic, financial, managerial and political context where the services are installed and operated.

Energy needs and consumption in developing countries

Energy consumption is recognized as an important indicator of development. The higher the average energy consumption per capita, the more developed the country (OLADE and World Bank, 1991). For example, the average per capita electricity consumption in South Asian countries was only 324 kWh/ year, compared with 8,238 kWh/year in higher-income countries (25 times greater) (World Bank, 2001). In specific instances, the gap is even greater. For example, in 1997 per capita electricity consumption in Canada was 15,829 kWh compared with 714 kWh in China and 181 kWh in Cameroon.

Research on rural electrification suggests that rural people without access to electricity only need small amounts of energy to improve their quality of life and their income. In most cases, electricity is necessary for the provision of lighting and community services such as education and health. Activities for transforming or preserving products (such as milling grain, chilling milk or making ice) are performed at a very small scale, thus requiring low amounts of energy.

There is evidence that in rural areas in developing countries, energy consumption rarely exceeds 30–50 kWh per month. Practical Action's experience in Peru shows that in isolated rural communities in the Andes, for about 60 per cent of the population, consumption barely exceeds 30 kWh per month per household (Sánchez Campos, 2005). In Thailand, during the process of rural electrification, the average energy consumption during the first five years was 11–22 kWh per month. After the five years, this figure rose to 22–50 kWh per month for certain villages (Vorvate and Barnes, 2000).

Another important factor affecting demand for electricity in rural areas is the dispersion of consumers as well as the subsistence production activities they undertake. Therefore not only are energy requirements for economic production very small, but also distances reduce consumers' incentive to take their products to market, thereby keeping their energy needs low.

Rural energy options

In the past electricity used to be supplied from small-scale generation plants. However, rapid urbanization, industrial development, the rapid development of technologies for electricity generation and economies of scale have meant that the world's electricity needs are now supplied through large grids. Unfortunately most rural people cannot access these grids.

Rural energy needs can be met through small stand-alone energy schemes. Appropriate options include renewable energy schemes: solar photovoltaic (PV) systems, small-scale wind electricity generators, small-scale hydro schemes and biomass systems. They are considered appropriate because they use local resources, can be sized according to need, operated and managed locally, and because local people can participate in the processes of planning and installation. Diesel sets have frequently been used for applications of this kind and can still be useful, but only when the fuel supply is reliable and affordable. However, this option is generally limited to few hours of supply because of the high cost of diesel fuel.

Renewable energy technologies were primarily and initially developed in the face of the energy crisis (in the late 1960s and early 1970s). From the 1980s onwards, they were also seen as the appropriate alternative solutions to protect the planet from the greenhouse effect. Despite their high cost in most cases, renewable energy technologies have progressed significantly. Between 1995 and 2001, wind installed capacity rose from about 5,000 MW to 28,000 MW. Photovoltaic capacity increased in the same period from about 0.1 MWp to about 1.5 MWp. In both cases there was a yearly growth in capacity of around 25 per cent during the 1990s.

This progress has tended to be mostly within developed countries. In the developing countries, during the last decade the adoption of renewable energy technologies has been driven by the need to lower greenhouse gas emissions. However, there have been significant efforts to use stand-alone energy generators. In fact, by 2000, PV systems for lighting were reported to number 150,000 in Kenya, 100,000 in China, 85,000 in Zimbabwe, 60,000 in Indonesia and 40,000 in Mexico. Around 150,000 PV and wind systems have been used for health clinics, schools and other communal buildings worldwide. Over 45,000 small-scale hydro schemes are operational in China, providing electricity to more than 50 million people. Over 100,000 families in Vietnam use very small water turbines to generate electricity from hydro schemes, while more than 50,000 small-scale wind turbines provide electricity in remote rural areas in the world.

Rural electrification in Latin America

Structural reforms in the energy sector in Latin America

The energy crises of the late 1960s and early 1970s had a profound influence upon the energy agenda in Latin America, and among the important initiatives taken was the creation of the Latin American Organization for Energy Development (OLADE[1]) to oversee energy policy for the region. OLADE promoted renewable energies during the 1970s and 1980s and built regional capacity to assess energy needs, and design and implement projects involving renewable energies.

The countries of Latin America looked to other regions to learn how to reform their energy sectors and the first to meet the challenge was Chile in 1982. The 1990s saw the major wave of reform, with Argentina, Peru, Bolivia,

Colombia, Guatemala, Panama, El Salvador and Nicaragua all reforming their energy sectors in that decade. The private sector responded positively with high levels of foreign investment and there was a consequent increase of electricity generation, transmission and distribution assets. In most countries, this caused a significant improvement in electricity coverage. The total installed electricity capacity in Latin America rose from 39,159 MW in 1970 to 92,578 MW in 1980, 120,713 MW in 1985 and 154,023 MW in 1988 (OLADE, 1991). Within Latin America there is a pronounced disparity between countries in terms of rural access to electricity. In Argentina, Mexico, Chile and Costa Rica, around 95 per cent of rural people had access to electricity in the late 1990s (OLADE, 1999), while in the countries of Honduras, Bolivia and Peru only 20–30 per cent of rural people had access to electricity.

Statistics show that rural electrification in Latin America increased significantly in the 1970s and 1980s, faster than in other regions in the world (see Table 1.1). Nevertheless, by 1990, the number of rural people without access to electricity exceeded 76 million. It has been argued that the structural reforms that occurred in the 1990s did not make a positive impact on the rural electrification sector because of the limited potential for profit-making within the rural context. For this reason, the widespread electrification of rural areas has long been postponed.

There are some examples of specific strategies and legal frameworks to promote rural electrification that have been in place for a number of years, for example the rural electrification concession in Jujuy (north-west Argentina) and the Renewal Energies Programme (Programa de Energías Ronovables – PER) in Chile. Nevertheless, it is only now that most countries within Latin America are developing and adjusting to new legislation to promote rural electrification. They are beginning to develop rural electrification strategies, pilot programmes and projects, mostly with the support of international agencies such as the World Bank, United Nations Development Programme (UNDP), Inter-American Development Bank (IDB) and others. In all cases the new strategies and reforms consider it vital that governments subsidize the development of the rural electricity sector.

Stand-alone schemes for rural electrification

In Latin America any extension of electricity provision to rural areas over the last few decades has been achieved through grid extension, and to a lesser extent though diesel sets. Comparatively few people have come to acquire electricity generated by renewable sources of energy, such as small-scale hydro, solar PV and wind energy. These have only attained the status of pilot programmes.

From late 1970s there has been an important effort to promote renewable energies in the region. Among the most active organizations involved in this drive have been OLADE, the National Aerospace Institute in Argentina, Institute of Research on Industrial Technology and Technical Norms (ITINTEC) in Peru, the National Institute of Electricity in Mexico and ELECTROBRAS in Brazil. Important international agencies included GTZ, which implemented the micro-hydro energy programmes, Special Programme for Coastal Atlantic Energy (PESENCA) in Colombia, Hydroelectric Microcentral Development Programme (PROMIDEHC) in Peru and the Programme for the Promotion of Renewable Energies (PROPER) in Bolivia. Practical Action has also been operating its energy programme in Latin America since 1984, primarily in Peru but also in other countries, to promote renewable energies for the benefit of the rural poor.

Nevertheless, renewable energies are still not a favoured technology for rural electrification. Connection to the grid remains the first choice, despite high costs. For example, in Brazil, connection to the grid costs between US$1,000 and US$5,000 per household for small clusters of 20–60 rural households at distances of 2–10 km from the grid. In Peru, connection to the grid costs more than US$1,300 per family for those villages next to the grid, and it is likely to remain extremely high for many small and isolated villages.

There are examples of efforts to introduce more stand-alone schemes for rural electrification, such as the ELECTROBRAS BRA/00/015 project, which will provide electricity services to one million households (five million people) through the installation of solar PV in the Brazilian states of Bahia, Caerá and Minas Gerais. In Peru an initiative by the United Nations Development Programme Global Environment Facility (UNDP-GEF) for the installation of 12,000 units of 50 Wp solar PV systems began in 2000. Guatemala has a project to install approximately 10 small-scale hydro schemes, and other projects are planned for Nicaragua, Ecuador and Bolivia.

Despite these efforts, Latin America is still far from overcoming the challenge of achieving global access to electricity; this challenge needs more commitment of governments' financial resources, but the primary and more urgent need is for reform of the energy sector and the development and implementation of the right strategies to tackle the barriers to rural electrification.

The electricity sector in Peru

Electricity coverage

By the end of 2005, 73 per cent of the population of Peru had access to electricity. Access is uneven, however, and the better-served areas are those located along the coast of the Pacific Ocean, followed by those in the rainforest region. Those areas in the Andes region have the least access. However, when considering electrification of rural areas alone, the figures are different: rural access to electricity is worst in the rainforest region, followed by the Andes, then the coast. The key reasons for these disparities are:

• Proximity to the national grid: regions located along the Pacific coast have benefited the most because the grid follows the coast to embrace the cities of Lima, Arequipa, Moquegua and Tacna.

- Urbanization: regions with high urban populations, such as Lima or Ica, have greater coverage because the electrification policy aims to provide electricity to urban centres first.
- Centralization: historically most of the commercial and industrial activities of Peru were located along the coast, allowing easy transport and export by sea.
- Access to transport: poor infrastructure and a lack of roads affect the decision to bring services to communities. In both the Amazon and the Andes there are small, extremely isolated communities, some of which require days of travel by boat (in the Amazon) or by horse (in the Andes).
- Government policy: there have been no clear policies for rural electrification in the past. Only now is the government making efforts to design the appropriate policies and strategies. Up until now, the approach has been to invest in grid extension, focusing increasingly on more expensive connections (in less-densely populated or geographically difficult areas).

Despite the policies of the 1970s and 1980s to promote rural electrification and improve access, Table 1.2 shows that there was no significant change during that period. It was only in the 1990s that rural electrification showed improvement; as a result, the rural electricity coverage of Peru is still very low when compared with other countries in Latin America.

Stand-alone schemes for rural electrification in Peru

A large proportion of access to electricity in rural areas is provided through the national grid. However, stand-alone schemes (diesel and small-scale hydro) do contribute to rural electrification. According to official statistics from the Ministry of Energy and Mines in Peru (Ministério de Energía y Minas del Peru, 1999), by the end of 2002 there were nearly 200 diesel sets of less than 1 MW (of these, 64 were below 100 kW) and nearly 150 hydro schemes of less than 1 MW (of these, 38 were below 100 kW). These figures do not include another hundred or so hydro units, all below 100 kW power, installed by international agencies, local authorities and communities, or similar small-scale diesel schemes. The statistics demonstrate that isolated schemes can play a role in rural electrification in Peru. However, progress in implementing projects of this kind has always been slow.

Table 1.2 Progress of rural electrification in Peru

Year	Rural electrification (%)
1972	2.5
1992	3
1995	5
2000	25

Source: World Bank (1999)

There are two main issues that account for the lack of interest among governments and other agencies in promoting stand-alone schemes for rural electrification: sustainability and cost. It is recognized that the high costs of stand-alone schemes could be overcome either by implementing programmes with more subsidies and/or building local capacity. There is, however, a need for further research into the sustainability of stand-alone schemes through the establishment of pilot projects and the study of ongoing schemes. Sustainability issues present a real barrier to progress. Even when governments and leaders may be willing to provide electricity services to marginalized communities, there are obstacles that prevent rural populations from accessing energy, especially those from isolated communities who have no other options but small-scale stand-alone energy schemes.

During the last three decades, international agencies as well as national research institutions have made important contributions to building local and/ or national capacity through projects and programmes of technology development and transfer, training and piloting schemes. As a result there are skilled, local personnel capable of designing, manufacturing and installing renewable energy schemes in a number of developing countries. This has resulted in lower initial investment costs through cheaper equipment manufactured in the country concerned, cheaper materials and cheaper construction processes. Nevertheless, their sustainability when installed in isolated rural areas is still in question. Despite some good examples, the poor technical, managerial and financial performance of the majority of isolated energy schemes is still of great concern to planners and decision-makers.

Table 1.3 The critical factors for sustainable stand-alone energy schemes

a) Critical factors in the local (community) context

Social	Financial	Technological	Managerial
• Benefits from electricity on household	• Low connection cost	• Local capacity	• Management model
• No political interference	• Effective bill collection	• Load factor/end uses	
• Community participation	• Capacity to pay	• Technical support	
	• Cost of energy	• Source of energy	

b) Critical factors in the wider context

The degree of development of the village

Market connections

National capacity

Legal framework

Lack of bureaucracy

Proper tax framework

Source: Sánchez Campos (2005)

It is therefore highly important to identify the critical factors that account for the success of stand-alone energy schemes in remote rural areas. There is a need for new implementation approaches, new methodologies and models for operation and maintenance, strengthening and/or building local capacity, as well as more work on promoting productive uses of energy for income generation.

The literature on rural electrification highlights certain critical factors that influence the success or failure of stand-alone schemes; these can be grouped into social, financial, technological and organizational factors (see Table 1.3). The work of Practical Action in Peru suggests that factors emanating at the community level are more important for success or failure than those that pertain to the national or regional context. These local issues also require a great deal of effort and time to overcome; those of national scope are important but essentially they depend on the willingness of the government to implement appropriate reforms.

CHAPTER TWO

The management model

Management services for small-scale off-grid electricity schemes

Although there are several critical factors that determine the success of stand-alone electricity schemes, through installing schemes in different social and economic contexts and through researching the results of different small-scale rural electrification schemes, Practical Action has learned that the lack of local capacity to manage the schemes efficiently is one of the most important barriers to success.

A post-implementation study of performance in off-grid generating schemes, installed in Peru during an ESMAP/Practical Action project between 1996 and 1998, shows that small-scale off-grid schemes supplying electrical energy, including those managed by the state, exhibit serious difficulties in achieving sustainability. The study went on to design a management model to fit with rural conditions. This model became known as 'management services for small-scale off-grid electricity schemes'. The model attempts to embrace all the local village-level issues that affect sustainability (see Chapter 1). The model was initially piloted in the district of Conchán, in the province of Chota, department of Cajamarca (see Plate 2.1). It has since been implemented in other areas by Practical Action and is now being adapted and employed in other contexts.

Objectives of the model

The principal objective of the model is the *efficient management* of small-scale stand-alone electricity schemes, where 'electricity scheme' is taken to mean all the physical elements related to the generation and distribution of electricity: energy resources, civil infrastructure, electro-mechanical equipment and the electricity transmission (when needed), distribution networks and household connections. 'Management' is taken to mean all the activities involved in sustainably operating, maintaining and administrating small-scale electrical schemes.

The model introduces the concept of entrepreneurial management, which is generally new in remote rural areas where, historically, management models for electricity services have been run by central governments, municipalities, communities or cooperatives and in very few cases privately managed by the service owner.

The model seeks to promote financial management free from the interference of political interests, be they from within the community or outside, with the

Plate 2.1 The pilot project in Conchán
Source: Practical Action, Peru

appropriate number of people for operating, maintaining and managing the service; this has positive consequences for the financial performance of the scheme, as it is viewed as a business or service. It also aims to establish a culture of paying for the electricity services, attempting to do so in the fairest way possible for the different user types (payment must be in accordance to consumption). It is difficult to generalize about tariffs and what these should cover in poor and isolated communities, however, the income of the scheme must cover at least its operation and maintenance costs.

The model promotes and supports the rational use of energy and its use for income generation through the provision of local services and the transformation of products, thereby seeking to promote community development. The participation of the population is promoted as much as possible in the planning of the management system and in decision-making about the operation, maintenance and management of the scheme. In this way, people assume their responsibility for the scheme and recognize the rights and obligations of the different actors.

The model builds on existing local human capacity to develop skills in managing small-scale schemes efficiently. It takes into account internal relations, values and principles and creates or empowers local capacity to ensure the sustainability of the scheme. The core idea is that the enterprise must always belong to the local area.

Actors

There are four main actors involved in the operation of stand-alone electricity schemes: the owner, the micro-enterprise operating the service, the users (households that receive the service) and the users' auditing committee. Each of these has a clear and well-defined role in the model, as illustrated in Figure 2.1. They have assigned responsibilities and established rights in relation to the role they play. For example, the owner must respect the contracts signed with the enterprise, while the enterprise must fulfil its responsibilities for operation, maintenance and management; in exchange, it receives financial remuneration for its efforts.

A set of mechanisms has been developed to ensure responsibilities are fulfilled. The rules are established in specific documents and agreed to by all parties. All the actors are aware of the contents of each document. These are developed within the current legal framework. Legal tools and means are employed to enforce the agreements; for example, contracts are registered with notaries.

The implementation of the model requires a long consultation process with the community on the viability and sustainability of the scheme. Extensive training is provided on the different issues inherent in maintenance, operation and administration, service costs, lifetime of the scheme and other related issues. Agreements arise out of intense dialogue about the role of the actors, the operating enterprise and the instruments for applying the model. Once agreements have been reached, these are formalized and the actors receive their formal set of responsibilities in writing, which gives them the force of

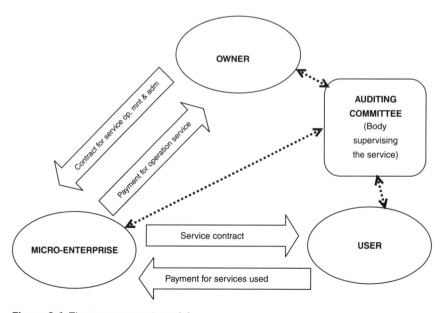

Figure 2.1 The management model

law. Essentially, the community gains a sense of responsibility, commitment, duties and rights, without losing the communal sense of ownership and cooperation among themselves.

The model has been piloted since 1999 in the micro-hydroelectric scheme of Conchán. To date five more schemes have been organized in the same way. The achievements show that this concept has been a success and that it is applicable without limits of scale, to pico-, micro- and mini-electricity generation schemes used to supply energy to small communities. These can be hydraulic or diesel, but the tariff structures applied will differ in accordance with the cost.

Relationships between the actors

The owner: To apply the model correctly, it does not matter who the owner is but it is important that ownership is clearly defined. In practice, it could be the community, a state agency, the church, the municipality or another agent; what matters is that there be an owner both capable of enforcing their property rights under all circumstances and also aware of the responsibilities they must comply with.

In Peru, in the case of electricity services for small rural towns, the body that raises the funds for implementation of the scheme tends to take ownership. When a municipality raises the funds,[1] it will claim ownership automatically, although in the majority of cases, it is not equipped to manage such schemes effectively. In real terms, a property owned by the municipality belongs to the people; however, mayors will claim the schemes as the property of their municipality and in most cases use them as a tool to strengthen their power. In other cases (communal, private or cooperative ownership), the ownership situation is clearer and well defined, with each relevant agent being the owner of the scheme. The community exercises their ownership rights through their representatives, as in the case of the cooperative. In the case of private ownership, the investor has the right to property, whether they are an individual or a legal entity.

When the government energy sector has installed a scheme (generally the Ministry of Energy and Mines – MEM), the owner tends to be the government,[2] and the scheme is incorporated into the government's rural electrification infrastructure, previously run by ElectroPeru, now by ADINELSA (Empresa Administradora de Infraestructura Eléctrica SA). However, when the scheme is installed by central government agencies outside the MEM, ownership is not generally clearly defined. Common practice is that the agency hands over the management to the community, represented by an electrification committee elected by the community. However, schemes are handed over without the agencies' giving any kind of training to the members of the electrification committee. Other agents like NGOs or churches generally give a certain amount of training to the organization. This training rarely goes beyond providing basic technical knowledge about operations and maintenance, and does not cover the administration or management of the service, nor other issues related to the social and economic implications to the community.

The model clearly addresses the issue of ownership and also clarifies the responsibilities of owners and how they can achieve the smooth running of the scheme. It defines the proprietor as the named owner of the infrastructure in relation to the other actors in the model and to the different institutions and/or individuals who interact in any way with the scheme and actors. The owner is responsible for the efficient management of the scheme, overseeing the provision of an electricity service of the highest possible quality, taking into consideration the capacity and limitations of the scheme. The owner also signs contracts with the enterprise running the electricity services, overseeing adherence to the contracts and the services, and takes part or is represented in the auditing committee.

The small enterprise: Under this scenario, a local (private) micro-enterprise takes charge of the management of the scheme. This enterprise is selected through an open and public tendering process, giving equal opportunities to all the local inhabitants who believe themselves capable of assuming the management of the electricity service and are keen to do so.

Selecting the local private micro-enterprise is complicated and requires a great deal of work, since in rural communities such enterprises normally do not exist. In most cases there is not even local capacity to manage a micro-enterprise. Therefore an ad hoc approach is employed by the model. This was designed for this process and tested in all the cases piloted to date. First the terms of reference are prepared, then all the people interested in operating the scheme are invited to compete under equal conditions; that is, applications are invited from all who believe they are capable of operating the service, with educational talks and simple information leaflets provided. Before the installation, it is important to assess local capacity to highlight needs and the level of training necessary before the recruitment process.

For the recruitment process itself, the minimum levels of achievement for candidates are identified, the selection criteria are agreed with local authorities and leaders and the selection procedure begins. An evaluation committee is formed, which defines the evaluation criteria and selects the enterprise. Once selected, the winning enterprise completes the necessary procedures to register itself as a legal entity and then takes over the management of the electricity services. Once the enterprise is chosen, it receives all the necessary support to complete its registration with the corresponding government agency.

One advantage of this model is that it brings down costs for running electricity services and supports the efficient and sustainable management of the scheme; it also clearly contributes to the creation or strengthening of local capacity and promotes the concept of enterprise (in some cases these may be the first enterprises formed in the area).

The main responsibility of the small enterprise is to manage the service efficiently. This includes energy supply, billing, bill collection, connection to new customers, disconnections and reconnections when required, and all the activities related to the operation and maintenance of civil works, equipment and electrical wiring of the scheme. The responsibilities of the operating small enterprise are detailed in the contracts.

The small enterprise trains the users (clients) on the rules, tariffs and the practical ways to make reasonable and efficient use of electrical energy. At the same time it trains the local population about the sustainability of the scheme. It signs contracts with each user, using a standard form developed for this purpose, in which it undertakes to meet good service standards. It also contributes one member to the auditing committee.

The users: The users are the people (the customers) who use the electricity service provided by the scheme and pay for it. To obtain the right to the service, they need to request it and accept a supply connection via electric cables. This service can be for domestic purposes, small businesses, institutional requirements or other needs. Each of the users is considered to be a client, and receives the rules for energy use and commits to adhering to them in a contract with the operating enterprise.

The auditing committee: The fundamental role of the auditing committee is to supervise service delivery. It is made up of representatives from the owner, the small enterprise and the users. The users will elect a representative in an open assembly, then nominate her/him to be part of the auditing committee. This committee controls and inspects as necessary to ensure responsibilities and obligations are met by each of the three other actors in the management model. It will solve claims and conflicts within parties when they happen. It must be impartial and its supervisory responsibility must be completely devoid of political interests.

The auditing committee performs its supervisory function either under its own initiative or in response to complaints from users about poor service or other issues. It generally does not discipline actors, but the representative of the community will clearly explain the actions of the committee to regular assemblies of the community. Any complaint is aired in the committee and agreements must be communicated to the interested party.

Operation of the model

The operation of the electricity service and the relationships between the different actors are clearly defined in the code of conduct on operation and functions. The contracts establish the rules and the operating mechanisms to run the service effectively. As such, the contract defines the principal functions, responsibilities and rights in the relationships between actors.

The owner of the scheme hires the micro-enterprise to manage and operate the scheme as a whole, that is, to carry out the necessary tasks to ensure the day-to-day functioning of the scheme according to the agreements. The relationship between the owner and the micro-enterprise is governed through a contract developed and agreed by both parties, signed and publicly registered. This contract documents the responsibilities, rights and duties of each party, and is valid for at least five years. In exchange for the enterprise's efforts, the proprietor pays it a monthly sum. The rules on collecting payment from users and the use of this money are established as part of the tariff structure (see Part II, Chapter 4). The amount to be paid is defined and agreed upon and fixed in the contracts, in order that both parties adhere to it. The contract also fixes the rules on disputes and considers how these can be resolved, and it takes into account relevant local law codes where necessary. The total paid for the service must under no circumstances exceed the sum collected from billing. Generally the payment to the micro-enterprise is calculated according to the effort required to manage the scheme. It is recommended that the structure of the payment be composed of two parts of a fixed amount (a base payment), plus a variable amount, which depends on performance in bill collection. This arrangement is an effective incentive for the enterprise not to tolerate late payments from users, to promote the electricity service to new users, and even to save energy, because in this way they can reach new customers and get more income.

The direct relationship between the enterprise and the user is simple – the user requests the service, accepts the tariff and signs a service contract with the enterprise (because the tariff has been agreed before in community meetings). The contract covers the liabilities and penalties of the enterprise while users are also bound by the contract and service terms. The electricity service contracts have a single format and are applied equally to all users, irrespective of their condition: they do not privilege any users. The relationship between the enterprise and users is also governed by a code of conduct, which is developed before or immediately after the scheme or management model is launched for the approval of the community and authorities.

The enterprise provides the service and all the relevant information to users on the scheme conditions and operations. It sends monthly bills and collects payments. It will also provide assistance to install new connections, extensions of the services and other services required at household level.

As outlined in the previous section, the user committee or auditing committee primarily acts as a trustee overseeing the quality of the service. As such, it does not have any written documents with the users, but it must respond to requests from any of the actors.

In contrast, there are direct contracts governing the relationship between the owner of the scheme and its users. These appear in the code of conduct and both actors are thus made aware of their responsibilities and those of the other actors. They show each other mutual respect and hold discussions when necessary. The proprietor does not have any powers to intercede on behalf of any user in the case of negligence or abuse by the enterprise. The owner is completely neutral in disputes between any of the actors.

Applications of the model

The model has been successfully piloted in stand-alone micro-hydroelectric schemes. The tariff system designed, tested and recommended is that of descending blocks. The results have been very successful in each of the cases, and despite the newness of the model it has sometimes overturned old styles of community management.

Practical Action found that the management model could be easily and conveniently applied to other decentralized energy schemes, such as solar, wind-powered, biomass or diesel sets, provided these services were implemented using a mini-grid (as in the case of hydro). To achieve this broader application, the definition of ownership needs to be modified to resemble the case of hydro and there needs to be a single owner who can then perform the function of owner as indicated in the model. However, for cases such as solar PV, where each user has an independent generating unit, a legal representative of all users is needed so that the enterprise can deal with one person or institution.

From the results obtained within the Peruvian context, it is clear that the management model can also be applied to electricity services from the grid when sold 'in bar' – an approach used by utilities to sell electricity in blocks[3] to a whole community (with one meter), and where the community manages the electricity distribution to its members and pays the bill in one block.

Instruments for applying the model

The model is simple in terms of its conception and design, but its success is determined by the appropriate use of a series of instruments. Some of these may be difficult to apply because they have implications for local economic development, and affect myths, beliefs and prejudices and the bad practices that can exist in any area.

From experience it is clear that one of the most important issues for the success of the model is the design and application of a fair tariff structure. Evaluations in the field of various cases suggest that a fair tariff is always based on what is consumed, applying a system of cost per unit consumed (see Part II, Chapter 4). Whoever consumes more pays more, therefore energy consumption must be metered regardless of the poverty or wealth of the community.

Contracts are also an essential instrument for the application of the model. They specify the commitments made by each party (owner, enterprise and users) within the current legal framework to ensure those commitments are met. In general, rural people are generally responsible and committed to meet any written contract they have.

The electricity service also requires a code of conduct on the rights and duties of the users. This encourages a regulated and respectful use of the electricity service and guards against mistakes by the users or the enterprise. It also provides avenues through which to resolve conflicts between the different actors.

As argued above, in rural areas generally there is little local capacity and few enterprises that can take charge of the scheme. Therefore the application of the model requires building that capacity through training in operations, maintenance and management for all those involved in the service (owner, enterprise and users). It is important to provide not only extensive training to the enterprise, but also sufficient awareness to the population about the common issues of this sort of electricity service: the need to pay, the lifespan of the scheme, safety aspects and the opportunities they have to use electricity for local development.

The participation of the local population in decision-making about the electricity service and the scheme in general must be organized. This is done through the auditing committee that enables users to have a voice. People can also participate through other sorts of organizations such as neighbourhood committees, which normally exist already.

Limitations and problems in operating or applying the model

From the cases where the model has been applied, certain problems or limitations are evident. First, sometimes the standards of service in isolated communities cannot meet the national standards regulated by the government, or it may not be appropriate or cost effective to meet such standards. Problems include voltage losses, type of equipment, location of equipment and so on. This gives ground to the national supervisory body to challenge or sometimes intimidate the community by demanding that they meet regulated standards. Such challenges may occur very few times in the life of the scheme, for example when an inspector of the national regulatory body visits them. Second, as mentioned above, where micro-enterprises are formed to manage the scheme, various limitations might arise owing to poverty and lack of knowledge and entrepreneurial initiative in rural areas. However, this can be overcome by flexibility, for example accepting individuals or groups who express a wish to compete to provide this service and establishing the micro-enterprise as a legal entity after its formation.

Third, the establishment of the micro-enterprise within an existing legal framework complicates the management task because this imposes taxation rules, regardless of the billing amounts. This means that although the enterprise may never or almost never need to pay tax on the small sums collected, they must fill in the complicated forms and regularly declare their income to the agency responsible for national tax collection. While this may appear to be an argument in favour of the informality of micro-enterprises, their legal status is essential in order that the contracts the actors sign be legally binding.

Fourth, the local population is strongly influenced by prejudices that go back several years. One of these is a 'culture of non-payment' – most times based on the belief that electricity is a service that should be provided by the state free of charge. This belief is often encouraged by irresponsible politicians who associate free access with the concept of social rights. With promises of securing free electricity, such politicians aim to win votes.

Allied to this is a fifth problem for it is still not possible to control political interference by local authorities in the management of the schemes, be this by the mayor, governor or others. Unfortunately this matter is outside of local control and it calls for the government to introduce an appropriate legal framework that tackles these sorts of issues to ensure the sustainability of the service.

Due to the remoteness of these services, disputes between the actors must be resolved through the participation of the auditing committee. The absence of other supervisory or advisory boards creates a final problem because it endangers the resolution of conflicts. Where the auditing committee proves incapable of resolving a dispute, there is no other authority where differences can be aired. It is hoped that once central government powers are decentralized to the regions, regional bodies managing energy and mines can undertake this role.

The instruments for applying the model were developed with the support of legal professionals. Their use is unavoidable because they determine the relationships between the actors. Through them responsibilities are assigned to each party. With the exception of the tariff structure (which should be substantially changed to suit each context), the other instruments can be used in other projects with few alterations to organize electricity and other services. The tariff structure, however, is appropriate to micro-hydroelectric schemes, particularly because it supports the promotion of productive uses and small businesses.

CHAPTER THREE
The Conchán pilot project

Introduction

The first piloting of the model took place in the district of Conchán in the province of Chota, department of Cajamarca, in the north Andes of Peru (see Maps 3.1 and 3.2). The pilot was part of an ESMAP-Practical Action project undertaken between 1997 and 1999. When the implementation of the model started in Conchán the population was around 870 households in approximately 170 dwellings.

The model was applied to a new electrical energy generation scheme that was an 80 kW micro-hydroelectric project built under the initiative and leadership of the local municipality in 1995–1996. The funding was obtained from the Ministry of Agriculture through the National Programme for Managing Hydrographic River Basins (PRONAMACHCS)[1] as a loan for US$128,000, payable over five years, with a 12 per cent interest rate.

The PRONAMACHCS loan to install the hydro scheme in Conchán was issued in 1994 and the project was completed in 1996. The municipality then began to make its repayments according to the contract with the lending body, which made clear that all of the loan would have to be repaid; however, by 1998 the debt had been cancelled.

The piloting of the management model was implemented in Conchán for various reasons. These include the fact that the hydro scheme was solely for a rural area. In addition, the mayor at the time had a good relationship with Practical Action and was aware that the management of the scheme was complex and needed to be self-sustaining. He also accepted the responsibility of paying off the contractual debt prior to its cancellation.

Prior to the implementation of the management model, a social, technical and administrative evaluation was conducted, which also evaluated energy consumption behaviours. The evaluation found that, until early 1999, the electricity service in Conchán was managed by an electrification committee, with the participation of the municipality. This committee was established for a period of five years (the contractual loan period agreed by the municipality), after which the municipality would be the owner and sole manager of the scheme. Although the committee was meant to be formed of a mixed group of local authority representatives and citizens, in practice, all the members were from the municipality. This meant that they mostly spent time on municipal functions, leaving little time for planning the management of the hydro scheme.

Map 3.1 The province of Chota in Peru
Source: Instituto Geofísico del Peru

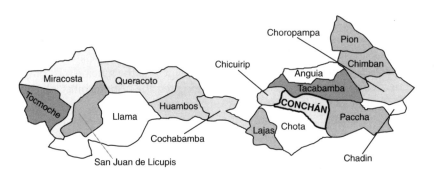

Map 3.2 The district of Conchán in Chota
Source: Instituto Geofísico del Peru

As a result, users of the electricity clearly identified the scheme with the municipality rather than with a representational committee.

Four people were primarily responsible for operating the scheme, namely two operators, a civil works technician and a bill collector who was responsible for invoicing (in practice, the municipal secretary). These four were contracted by the municipality of Conchán and received salaries for the management, operation and maintenance of the scheme that alone exceeded US$800 a month. In addition, there were six other people, staff of the municipality, who were involved in the management of the service. There was also an elected municipal councillor in charge of reviewing income, who could interfere in the decisions and even in the management activities of the scheme.

The monthly income of the scheme was estimated to be in the region of sol1,800 (US$540), coming from 120 households each paying sol15 (US$4.50). However, the real income from billing rarely exceeded sol1,200 (US$350) because payments were often delayed. The municipality had to cover the deficit to pay fees as well as the necessary costs of maintenance.

The operators each worked 12-hour shifts and there was always someone in the power house. These staff did not receive appropriate training, since the installing authorities believed that they only needed to learn about starting and shutting down the equipment when required.

There were no rules about the electricity service, the use of energy was indiscriminate and there were no checks, for example allowing the disconnecting of services for delayed payment, energy theft or other anomalies. During the Practical Action evaluation, there were significant contrasts between consumption patterns. Some families cooked on unreliable electric stoves, generally bought from the capital of the province or brought back from the coast. At the other extreme, there were families who used two 100 W bulbs for no more than four hours a day, two days a week. The electricity service as it was originally established did not take into account any of these varying consumption patterns and needs of residents in Conchán.

The municipality, represented by the mayor, was responsible for the management of the service. Therefore, if the mayor wanted to, the service could be used for political ends. The municipality was always under pressure to overlook late payments for the service, creating an additional pressure on the scheme.

The cost of an electricity connection was a one-off fee equivalent to US$28.50, a figure which had been determined in community assemblies by vote. Connections were made enthusiastically – everyone who approached the municipality was connected. There were no contracts for the service and there was no file with a list of users. Actually, the mayor had not given much thought to the management of the service and the complications involved. The quality of the service was poor. There were regular shutdowns and voltage variations, which in many cases fused domestic electrical equipment. The scheme did not (and still does not) correspond to the accepted technical specifications. Practical Action found that the project could not generate more than 40 kW of power, illustrating the makeshift approach and the poor supervision and technical support given by PRONAMACHCS. For example, the scheme had an inappropriate system of manual load control that caused some damage to electrical devices.

Implementing the management model

Once the implementation location was identified, Practical Action carried out intensive training for local authority personnel and community leaders in order that they understood the new model and agreed to its execution. Crucially, at each stage of the process, the role of Practical Action was to promote, facilitate and sometimes mediate, in order that decisions were always made by consensus between the interested parties, aiming for and reaching reasonable agreements regarding important issues, such as the costs per kWh for the different blocks of the tariff scheme suggested.

The process began with the definition of the ownership of the scheme, that is, who owned the assets. The process was complicated by the existing dispute between the municipality and the users. Finally, the scheme was declared to be the property of the municipality, since they had underwritten the loan and assumed the debt to the lender. Additionally, the municipality was using its own resources to repay the debt and subsidize operations and maintenance. These arguments were more than sufficient for the municipality to claim ownership.

Once the issue of ownership was resolved and the municipality had accepted the model, the next steps were to: present the model to the population and explain its advantages over the existing one; develop the tariff model through negotiation with the users; develop the code of conduct on the functioning of the service; and develop the other instruments to be used. In the case of Conchán, since there was already an existing management and tariff model, this phase was extremely delicate because it meant changing the rules and behaviours established.

Plate 3.1 Training the community
Source: Practical Action, Peru

The selection of a private local micro-enterprise culminated in the signing of the contract, in which the municipality conceded the operation, maintenance and management of the service to the Empresa de Servícios Eléctricos San Isidro (San Isidro Electricity Services Enterprise), created especially for this purpose by three associates. The service was conceded for a period of five years, after which the contract could be renewed or cancelled.

While the contracts were being negotiated, the local population was being trained about users' rights and obligations regarding the service, and on the different issues that can affect the sustainability of such a service. They were also trained on the need to have an appropriate tariff system that covers at least the costs of operations and maintenance, and enables savings for eventual repair costs. Table 3.1 shows the final tariff structure agreed.

To ensure the scheme worked smoothly, the technical problems had to be resolved, in particular the voltage regulation. With this aim the municipality, with Practical Action's technical assistance, installed an electronic load controller, to guarantee supply levels. Another basic requirement for the electricity service was to install meters. The municipality supplied funds so that all users could acquire meters, which were later paid for in 12 monthly instalments added to electricity bills.

As well as training, the staff of the San Isidro Electricity Services Enterprise were given basic instruments for managing the service, such as a computer, a printer, voltage and current meters and other tools, plus training in the use of these.

Results of the management model in Conchán

Practical Action monitored the impact of the implementation of the management model immediately after its application and subsequently in 2002

Table 3.1 The tariff structure applied in Conchán

Service charge	Units	Amount
Charge for energy consumption		
Up to 20 kWh	US$/kWh	0.15
Between 20 kWh and 60 kWh	US$/kWh	0.12
Over 60 kWh	US$/kWh	0.02
Other charges		
One-off connection charge for new users		
(charged at the time of connection)	US$	28.50
Monthly charge for street lighting	US$/month	0.28
Disconnection and reconnection fees (where these apply)	US$	0.57
Monthly interest on delayed payments as a percentage of the		
amount owed	%	2

Exchange rate: US$1.00 = S/.3.45, July 2001
Source: Sánchez Campos

Plate 3.2 Enterprise staff reading electricity meters in Conchán
Source: Practical Action, Peru

and in 2005. The results are impressive, owing largely to the micro-enterprise making constant efforts to reach new clients and improve income, without affecting users, and installing a small transmission line to a nearby village. The enterprise has also successfully managed the operation and maintenance of the scheme and repairs where necessary, resulting in the renewal of its contract for a further five years.

Every month, the San Isidro Enterprise receives a fee for service provision amounting to sol1,000 (US$300). It also receives an allowance, equivalent to 20 per cent of the difference between the income from the service and the service provision cost, as long as delayed payments remain under 6 per cent of the total. The enterprise keeps a register of users, contracts with each user for the electricity users, service request documents, invoices, accounts and a record of people registering, among other documents. All of these are part of the management process, and there is a software program so they can be computerized. As the San Isidro Enterprise developed and implemented its effective management system, the number of users of the electricity service has grown, as has the load on the scheme. In 1999, before the change to the management model, there were 110 users with a peak consumption of 26 kW. In 2002 there were 140 users with a peak consumption of 21.5 kW, and in 2005 there were 326 users with a peak consumption of 40kW. Figure 3.1 shows this evolution.

A high proportion of the users are paying an average of US$2.80 a month, that is, they are paying less than with the flat-rate tariff applied previously.[2] This is despite the fact that the minor consumers pay a higher tariff per kWh

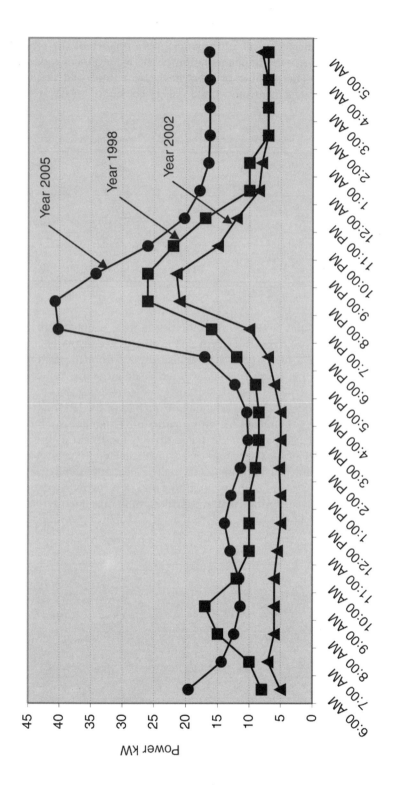

Figure 3.1 Evolution of load in Conchán

consumed. Overall, the technical and financial performance of the scheme has been very good since the implementation of the management model, as shown in Table 3.2. The number of users has grown three-fold since the application of the management model and energy consumption has not yet doubled.

Table 3.2 Technical and financial performance of the micro-hydro scheme in Conchán

Year	1998, 1999 (before model)	2002	2003	2005
No. of users	110	140	170	326
Monthly income	< sol1,000	sol1,500	sol1,700	sol2,800
No. of operators	4	3	3	3
Staff costs	sol1,800	sol1,070	sol1,140	sol1,340
Peak kW demand	25	21.5	30	40

Exchange rate: US$1.00 = S/.3.25

The tariff structure has awakened the initiative of local people to set up or improve small businesses. One case worth mentioning is the owner of a small business who sold ice and who has moved his business from the next town (Tacabamba) to Conchán because the energy tariffs are attractive and benefit his business.

Since being applied to the hydro scheme in Conchán in the late 1990s, the management model is working satisfactorily, maintaining high levels of income and low levels of delayed payments. It has a savings fund held in the local bank for replacements and maintenance that allows small contingencies to be resolved rapidly. The micro-enterprise managing the electricity services is now well known in the village and in the surrounding area. It is highly respected by both local population and municipal authorities, and consequently plays a full part in local civic life.

PART II
Instruments for the management model

CHAPTER FOUR
Tariff structure

The system

Choosing a tariff structure for a small-scale scheme delivering electricity services is complicated, to say the least. Projects to date have applied various structures including among others: flat-rate tariffs, which impose a single charge regardless of consumption; tariffs distinguished on the basis of use, which tend to differentiate domestic uses from commercial or productive uses; and state-regulated tariffs.

The system described below has been developed by Practical Action with the aim of complementing the management model described in Chapter 2. It is called the 'system of descending blocks'. The use of this tariff system requires the metering of energy consumption and the rates charged depend on consumption. It is formed of three blocks (see Figure 4.1). The first block corresponds to the first 'X' kWh, which coincides with the consumption of the sectors consuming the least energy in Peru; in isolated villages it is around 20 kWh per month. Households generally have two or three lights, a small radio and perhaps a small television. A second block corresponds to domestic consumers with some appliances such as a refrigerator, colour television and perhaps a small business (for example selling basic goods), and their consumption ranges between 20 to 60 and 70 kWh per month. The third block have high rates of consumption. These clients will be engaged in production or provision of some sort of service that consumes a lot of energy; among these are welding workshops, carpentry workshops, ice production, and so forth.

Projects can choose the number of blocks to use but, to facilitate management, working with three blocks is recommended. Block I is the block with the highest cost per kWh, Block II with middling costs and Block III with lowest cost. However, a larger number of blocks allows for a closer approximation to consumption and therefore a charge more closely linked to the client's consumption.

Once the ranges of these blocks have been set, prices can be simulated and the income derived can be examined. The income collected must be at least enough to pay for the operating and maintenance costs of the scheme and to save a small sum each month. This will go into a fund for corrective maintenance and unforeseen eventualities. For larger schemes, the tariffs can even be set at a level that allows for the repayment of the initial investment or for replacing the scheme at the end of its lifetime. In determining the tariffs, it helps to refer

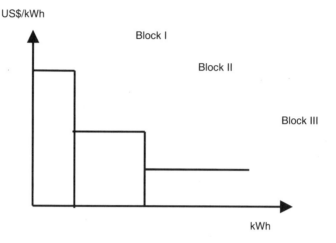

Figure 4.1 Tariff structure for the system of descending blocks

to some existing tariff values. In our case (in Peru), the official tariff for rural areas, BT-5, serves as a reference.[1]

Block I

In the cases implemented (Conchán, Tamborapa, Las Juntas, Huarango and Huarandoza), the energy consumed in kWh in Block I was assigned a cost similar to or slightly higher than the BT-5 tariff. However, it is recommended that the tariff be higher than BT-5, although the socio-economic conditions of the population also need to be taken into consideration as well as the estimated income of the scheme to keep a sound financial performance, since the model presumes that the plants will be self-sufficient. Previous experience shows that a high percentage of users is clustered in this block.

Block II

The cost per kWh in this block should be around the average tariff in rural areas (BT-5) or slightly below it. As mentioned above, this block comprises more intensive domestic uses because these users have electric appliances that consume more energy, such as refrigerators or televisions. However, the number of households in this block represents a low percentage of the population.

Block III

This block offers low rates. It is generally the block that includes small businesses, local services, production or manufacturing, and as such it is recommended that costs be kept low, as long as the income of the energy scheme is maintained at a reasonable level.

In addition to the blocks described, it is worth considering a minimum payment for the right to access energy. Connecting a household to the energy service that does not use it, or uses it minimally, can deprive other households. To avoid this type of situation and to improve the financial situation of the scheme, a minimum payment is recommended, which could be the cost of the first 8 to 10 kWh.

In Practical Action's experience, for all the cases where this model has been implemented, the take-up of energy for productive uses has been significant. In each case, a variety of small businesses, principally directed at meeting local needs, have sprung up, such as workshops repairing agricultural tools, welding workshops, tyre repair workshops, carpenters, restaurants and battery charging stations, amongst others.

Benefits of the system of descending blocks

The system has the primary aim of raising the highest income possible in order to make the electricity service financially viable, at least in terms of the operating and maintenance costs in the case of very small schemes (with fewer than 200 users). In the case of larger-scale schemes (over 200 users), this tariff model can even allow for the recovery of all or some of the initial investment.

The high cost per kWh in the first block improves the budget of the scheme, since the group of users with the lowest consumption rate is generally the majority of the population. In Peru and other parts of Latin America, about 60–70 per cent of the population consumes less than 20 kWh monthly.

Although the tariff scheme has different costs per unit according to the consumption level of the users, it tends to be as fair as possible because the users pay for what they consume in accordance with the tariff set. That is to say, higher consumers pay more, making the use of energy metering essential.

The households that fall in Block III normally use energy for productive ends. This is when the low cost of energy turns into an instrument to promote production. However, they also pay the same unit cost for the first kWh consumed. Experiences to date in different contexts where the model has been applied offer lessons. In Conchán, for example, a small ice-making factory from a neighbouring district was moved there. In Las Juntas, there was a similar case where a family with a small rice stacker moved. These moves were prompted by the favourable tariffs, since previously they operated under tariff schemes that did not promote energy to be used for production.

The low cost of the third block encourages its use for production. Additionally, this use does not conflict with peak hours, since generally productive uses take place during the day, while the peak demand is between 7 and 10 p.m. in the evenings.

It has been noted that households whose consumption falls within the first block generally spend lower amounts than those who do not have electric energy and thus buy candles and/or kerosene for lighting and batteries for radios. In fact studies in Peru show that those who do not have access to electricity expend US$3–10 per month per family and in some case even more. Therefore people can pay at least this amount or even more for better energy services.

The households whose consumption falls into the first block are not necessarily the poorest; some families are among those with the highest incomes; however, their intense agricultural activities force them to remain outside their homes for the majority of the time. The use of energy is principally at night then, for lighting, radio and TV and in some cases only at weekends. Therefore a minimum payment will always help to improve the income of the scheme and hence its sustainability.

CHAPTER FIVE

Code of conduct for the operation and functions of the electricity service

Introduction

In the case of Peru the terms and conditions of the scheme management need to be regulated within the following legal framework:
- D.L. No. 25844, LAW ON ELECTRICITY CONCESSIONS.
- LAW No. 27744, LAW ON RURAL ELECTRIFICATION, REMOTE AND OUT-LYING AREAS.
- LAW No. 27972, THE ORGANIC LAW ON MUNICIPALITIES.
- D.S. No. 009-1993-EM, BY-LAW OF THE LAW ON ELECTRICITY CONCESSIONS.
- R.M. No. 263-2001-EM/VME, BY-LAW ON HEALTH AND SAFETY AT WORK, IN THE SUB-SECTOR OF ELECTRICITY.

Given the above, the following guidelines must be adhered to:
- Any activities of generation, transmission and distribution of electrical energy that do not require concession or authorization can be freely carried out, but must be in accordance with national energy standards and environmental and cultural heritage regulations. However, the Ministry of Energy and Mines must be informed of the technical specifications of the installations at the start of the operations.
- The municipal council approves the concession of public municipal services through any private investment mechanism allowed by the Organic Law on Municipalities and laws on this issue.
- With regard to contracting – public tendering or other processes – municipalities are subject to the State Law On Contracting and Acquisitions (D.S. No. 012-2001-PCM) and its Code of Conduct (D.S. No. 013-2001-PCM), which requires adherence to the process in these laws and is subject to the supervision of CONSUCODE.[1]
- The municipal council has the power to approve the management practices for local public services.
- Municipal by-laws are general laws through which the regulation, management and supervision of public services are approved.
- The mayor has the power to supervise the smooth working and the economic and financial outcomes of municipal public utilities offered under delegation by the private sector.

- The provision of local public services is supervised by the municipal council and by the community, in accordance with the Organic Law on Municipalities.
- The processes for calculating tariffs and the maximum tariffs to apply will take into account the fact that there are no regulations for very small-scale energy generation, as is the case in most rural areas in Peru, unless the systems are installed by the government.

Example of a Code of Conduct

The Code of Conduct, like the other instruments, should be written case by case, taking into consideration the existing legal framework and people's views, this requiring the participation of people in the design of the instruments. This example should be used as a good base to start the process.

Heading I: General considerations

Article 1: In this Code of Conduct, when the terms 'The Law', 'The Code of Conduct', 'The Owner' (which would be the municipality or the community), 'The Operator' and 'The Service' are used, they are to be understood as follows:
- 'The Law' refers to the Law on Electricity Concessions.
- 'The Code of Conduct' refers to the Code of Conduct for the Operation and Functioning of the Electricity Service.
- 'The Owner' refers to the municipality or the community in the case of electrification of small district towns; it could refer to the state or a concession holder if the model is applied in such cases.
- 'The Operator' refers to the enterprise operating and managing the electricity service.
- 'The Service' refers to the public electricity service in the area where the Code of Conduct is applied.

Article 2: This Code of Conduct for the operation and functions of the electricity service in (name of place) has the objective of regulating the activities related to the generation, distribution and commercialization of electrical energy in (name of place), in the district of ...., in the province of ....., in the department of ....

Article 3: In accordance with the stipulations of Article 7 of the Law No. 25844, the Law on Electricity Concessions, referring to generation, transmission and distribution activities that do not require concession or authorization, such as Article 9, Subsections 18 and 32 of the Law No. 27972, the Organic Law on Municipalities, the District Municipality of .... (or the community or XX) has the power to manage and organize the public electricity service in (name of place), applying the organizational approach most conducive to the community's interests, the efficiency and efficacy of the service and a due level of municipal control.

Article 4: Having evaluated the advisability of having an efficient electricity service and taking into consideration the existence of a working generating scheme as a product of ... and having evaluated the different possible organizational alternatives, 'The Owner' has decided to hand over the management and operation of the public electricity service to a local enterprise, as documented in the Contract of Operation and Maintenance Services.

Article 5: The Public Electricity Service is offered by 'The Owner', through 'The Operator' by virtue of a Contract of Services signed by both, which permits 'The Operator' to offer the operation and management of the electricity service in (name of place).

Heading II: The electricity service

Article 6: The Electricity Service is defined as the right of the user to make use of the electrical energy within the limits and restrictions established in the Electricity Supply Contract, which the interested users must sign with 'The Operator'.

The electricity service is offered to a person or legal entity requesting the service (including for commercial uses), who acquires the status of User. The electricity service is offered at their premises.

Article 7: All the people or legal entities whose premises fall within the district limits may be considered as users if they meet the following criteria:
• being a legal adult;
• being able to prove that they reside in the premises as the owner or tenant.

Article 8: The electricity service can be offered to the users in three different types:
• single-phase electricity service at low tension;
• three-phase electricity service at low tension;
• three-phase electricity service at medium tension (when exiting).

Article 9: The electricity service is offered through an electrical connection, the spur line. The spur line is defined as the electrical connection between the distribution lines and a point inside the user's premises.

The energy supply service delivered by the operator ends at the point where the spur line begins. The maintenance of the spur line and the internal connections in the user's premises are the sole responsibility of the user, as is any occurrence caused by the state of repair of the spur line and the installations or inadequate manipulation of these.

Heading III: The functions and rights of the owner

Article 10: 'The Owner' is responsible for organizing the public electricity service. As such they will supervise its normal running and ensure that the conditions established in the Contract of Operating and Management Services and this Code of Conduct are met.

Article 11: To ensure the smooth running and the economic and financial results of the public electricity service, 'The Owner' will assemble a User Committee to appoint the Auditing Committee, which will be responsible for the direct supervision of the public electricity service offered by 'The Operator'.

'The Owner' will appoint a representative as a member of the Auditing Committee, who will be present as a full member.

Article 12: Every month 'The Owner' will receive a financial statement and a monthly report on ordinary occurrences from 'The Operator' and will be able to demand extraordinary reports when the circumstances call for this.

Article 13: 'The Owner' is responsible for paying the due amount at the times specified in the Service Contract signed by 'The Owner' and 'The Operator'.

Article 14: In the concession, under inventory 'The Owner' must provide 'The Operator' with the materials, spare parts and tools necessary for the activities it is committed to perform. The lack of materials, parts and/or tools necessary for service delivery exonerates 'The Operator' of responsibility in cases where its ability to perform its functions was hindered.

'The Operator' must maintain the materials, parts and tools provided by 'The Owner' in optimal condition. This excludes deterioration caused by everyday use.

'The Operator' is responsible for any loss or deterioration of these goods, be it through financial mismanagement, negligence or ill-use attributable to 'The Operator', the persons employed by it or those persons given access to the installations. In these cases, 'The Operator' must replace them or repay their value out of its own funds.

The conditions specific to the concession are stipulated in the Contract of Service Provision between 'The Operator' and 'The Owner'.

Article 15: As part of the concession, 'The Owner' will provide 'The Operator' with premises to be exclusively used as an office for the duration of the Contract of Service Provision, so that 'The Operator' can conduct its activities to deliver the electricity service.

The conditions specific to the concession are regulated in the Contract of Service Provision signed by 'The Operator' and 'The Owner'.

Comments

The concession contract must be signed at the same time as the contract of service provision and must contain the responsibilities of 'The Operator' and 'The Owner' with respect to the goods covered.

Repairs are not covered under these contracts; these should be assumed by 'The Owner'.

The concession is covered under the Organic Law on Municipalities and the laws on the management of fiscal property and the supervision of national goods.

Heading IV: The functions and powers of the operating and managing enterprise

Article 16: 'The Operator' is responsible for:

- The uninterrupted operation of the generating plant in accordance with user requirements, within the capacity of the plant and its operation, taking into account the points raised in Article 17.
- The commercialization of the electric energy produced, in accordance with the tariff provisions established in that case. There will be an established tariff agreed with 'The Owner'.
- The maintenance of the equipment, the installations and the energy production and commercialization infrastructure, in adequate working condition.

Article 17: In these activities, 'The Operator' is obliged to comply with the following:

- The electricity must reach the users at the agreed voltage and frequency levels.
- Interruptions to the service must be attributable to unforeseeable circumstances and not to the negligence of its personnel.
- Customer service must be polite, cordial and timely.
- 'The Owner' and/or the Auditing Committee must be granted access to all the information, documents and material related to the provision of electricity services. They must be able to visit and inspect the installations operated, be it directly or through nominated third parties.

Failure to comply with these obligations will incur the penalties set out in the Contract of Services between 'The Operator' and 'The Owner'.

Article 18: 'The Operator' is subject to the controls that 'The Owner' of the service and the Auditing Committee exercise under ordinary and extraordinary circumstances.

Heading V: Supervision of the service

Article 19: The supervision of the electricity service activities is the responsibility of the Auditing Committee. This committee is the highest authority approved by 'The Owner' to supervise, penalize and arbitrate in the case of disputes between any user on the Register of Users and 'The Operator' in situations solely arising out of the provision of the electricity service.

Notes

The Auditing Committee is approved by 'The Owner'.

This project does not distinguish between the terms 'Communal Assembly' and 'User Committee'. It is important to bear in mind that municipal law regulates various mechanisms of participation and local control such as:
• Open meetings of the town council, in accordance with regulatory by-laws.
• Participation through neighbourhood groups, committees, associations, communal or social organizations or similar local groups.
• Management committees.

Local groups are responsible for supervising the provision of local public utilities. They are created through municipal by-laws and must adhere to the laws stipulated in the corresponding Code of Conduct.

The voting mechanism and number of members of the Auditing Committee must be determined, as must the definition of an active member.

It must be clearly determined whether 'The Operator', 'The owner' or the users determine the tariff structure.

Article 20: The Auditing Committee is made up of five (5) active members, who perform the following functions:
• president;
• vice-president;
• secretary;
• two members.

In this committee, the users of the electricity service, 'The Operator' and 'The Owner' will be represented.

Note

It is necessary to define who elects the representatives. It is recommended that there be an odd number of members.

Article 21: The committee's responsibilities and powers are:
• To adhere to and ensure the adherence to this Code of Conduct regulating the operation of the electricity service.
• To ensure that the infrastructure for generating, commercializing and distributing electricity is maintained in good working order.
• To settle disputes between the users and 'The Operator', and penalize the party declared to be at fault. There will be a code of conduct regulating faults and penalties, clearly written and approved by committee, with the participation of the users, 'The Operator' and 'The Owner', in one of the User Committee meetings.

- To evaluate the monthly financial statements presented by 'The Operator' and to offer opinions on its management with the aim of eventually renewing its contract.
- To hold meetings once a month.

Article 22: The responsibilities of the president are:
- To act as the official representative of the Committee.
- To adhere to and ensure adherence to this Code of Conduct and uphold the Committee's rulings.
- To ensure that the other committee members perform their activities.
- To call and chair Auditing Committee meetings (regular and extraordinary).
  In the absence of the president, the vice-president will take their place, assuming the role and responsibilities of the president.

Article 23: The responsibilities of the secretary are:
- To keep an up-to-date record of the minutes, rulings and files of the Committee.
- To handle the internal and external correspondence and sign documents with the president.
- To keep other members informed of agreements and rulings of the Committee, keep attendance lists and keep note of the quorum, noting in the minutes the whole text of rulings and initiating the corresponding procedures.
- To call the meetings of the Auditing Committee, summoning third parties in the cases where this is required.

Article 24: The responsibilities of the members are:
- To participate actively in the regular and extraordinary Committee meetings.
- To participate in regular and extraordinary committee meetings.
- To participate in the commissions formed to investigate specific cases.
- To gather and channel the users' complaints.
- To disseminate to users the contents of this Code of Conduct and any alterations to it.

Article 25: The Auditing Committee meets regularly every 30 days and extraordinarily when a meeting is called by the president or requested by any of the members. The Committee can appoint special commissions to analyze or study specific cases, informing them of their specific aims and duration, which will be in accordance with their function.

Article 26: The Auditing Committee consists of ... (DEFINE COMPOSITION)..., with equal rights as full members.

Article 27: Regular meetings of the Committee will be conducted within five days of receiving the monthly report from 'The Operator' and, for each member, attendance at the meeting is absolutely compulsory.

Article 28: Extraordinary meetings will only be held to resolve important issues, which were not foreseen in the regular meetings.

The extraordinary meeting will take place when required by the president of the committee or when half plus one of the total members of the committee request it.

Article 29: The quorum for regular and extraordinary committee meetings is half plus one of the total of the members.

If the minimum for the quorum is not present, the president of the committee will postpone the meeting. The postponed meeting can be on the same day, at least one hour after the first. In the second meeting, the members who are present will make the rulings and the quorum will not be required.

Article 30: Participation in the Auditing Committee is ad honorem, since no material advantage is pursued.

However, the active members of the committee will be entitled to a 30 per cent discount on their respective monthly electricity bills.

The members of the committee who are stripped of their position will automatically lose the benefit of the 30 per cent discount.

Article 31: Criteria for membership of the committee are as follows:
• Being a legal adult.
• Not holding a public position that by its nature compromises the impartiality of the person.
• In the case of inhabitants from other areas, at least five years' continuous permanent residence in the area where the service is provided.

Article 32: The members of the committee will hold their positions for a period no longer than one year. Each of them can only be appointed once. Once the period is finished and an ex-member has been absent for at least one period, they can stand again and be elected for a new period, which is considered a totally new period.

Article 33: It is the responsibility of the users to pay the monthly bills received from 'The Operator'. If the user disagrees with the total payable or one of the figures on the bill, they must present their verbal complaint to 'The Operator', who must deal with the complaint immediately. If 'The Operator' does not uphold the complaint, the user is free to appeal to one of the members of the committee. The member can collect the complaint and can also request a report on the case from 'The Operator'.

The committee member who received the complaint, and who knows the position of both parties, will present it in the regular committee meeting, adding a personal opinion. After checking against this Code of Conduct, the contract and other relevant documents and, in keeping with previous rulings, the committee members will discuss the case exhaustively and then proceed to vote on it.

Once the result of the vote has been made public, the case will be considered closed. The result of the vote will be recorded in the minutes, as well as a brief report, which must be presented to the parties involved.

For the rulings of the Auditing Committee to be valid, these must have been agreed by at least half plus one of the members.

Article 34: The decisions or rulings of the committee are final. However, the party who consider themselves affected by the committee's decision are free to exercise their legal rights.

Article 35: The members' discussions within the committee must take place in a spirit of order, initiative, goodwill, impartiality and discipline, referring to this Code of Conduct as a framework.

Within the committee, it is absolutely prohibited to defend or promote any political, religious, racial or social ideology. To do so repeatedly is considered a grave fault in members.

Heading VI: Tariffs

Article 36: The prices for the sale of electrical energy, the various charges (for street lighting, disconnecting or reconnecting electricity supplies, arrears on delayed payments, connection costs) and the tariff for financial penalties for users (resale of energy, clandestine use) are established annually and take effect from the first of January of each year.

Article 37: The tariff is adjusted to the specifications established according to the system of descending blocks. This is firstly designed with the users and then agreed with the users. This model has as its principal objectives the financial viability of the model and the promotion of productive uses.

The results of the tariffs applied must be evaluated annually, taking into consideration the following:
• Projected annual average operation and maintenance costs for the generation, distribution and commercialization of electrical energy. The cost of materials, parts, tools and the payment to the operating enterprise must be taken into account.
• Generation capacity of the hydroelectric scheme in kWh per year.
• Projected income from the sale of energy.

Article 38: The application of the tariff is the responsibility of 'The Operator', while the committee is responsible for supervising its correct application.

Heading VII: Collecting payment, delayed payments and customer service

Article 39: Collecting payment is an activity corresponding to the commercialization of electrical energy. Therefore it is the responsibility of 'The Operator' and is subject to the supervision of the Auditing Committee.

'The Operator' will inform the users of the payment deadline with at least fifteen (15) days' notice, using the most effective communication medium.

It is the responsibility of the users to find out the deadline for payment. Therefore not knowing the deadline for payment will not be considered a valid reason for delayed payment for the energy consumed in a month.

Article 40: The deadline for payment determines the date until which the user can pay for the energy consumed without incurring additional charges.

After the deadline, the user will have to pay an additional charge for defaulted or delayed payment. Each year, 'The Owner' will set the rate for each additional day. This charge for late payment has the aim of:

- Covering the administrative costs generated by the delay in payment, since staff have to commit time to this activity instead of another service activity.
- Creating awareness amongst the users of the negative effect of delayed payment on the normal working of the electricity service.

Article 41: Opening hours for receiving payment for energy consumed, complaints, reports or queries, as well as working days, must be communicated by 'The Operator' using the most appropriate media.

Article 42: Customer service is a priority for 'The Operator'. Politeness, efficiency and timeliness of services to users are indicators of 'The Operator's' management of the electricity service. These will be considered by the municipality and the Committee when reviewing 'The Operator's' service contract for renewal.

Heading VIII: Disconnecting and reconnecting users

Article 43: When a specific user has failed to pay three months' worth of energy consumed, their electricity service will be disconnected.

This service will only be reconnected by 'The Operator' once the user has paid the debt and the additional charges for delayed payment, interest and administrative charges.

Article 44: The disconnection and reconnection of the service incur a cost, which must be met by the user as stipulated in the previous article.

'The Owner' will meet the costs of disconnection and reconnection incurred through the maintenance, renovation or improvements to the infrastructure for distributing electrical energy.

Heading IX: Invoicing and the register of users

Article 45: The administrational activity of invoicing is performed by 'The Operator' every month, to calculate the total amount payable by each user for the amount of energy consumed, street lighting etc. and to send out the invoices and prepare statistical information.

Invoicing includes the printing of the invoice for each user, the generation of statistical information about sales and production of electricity (in financial terms and in units of energy) and the payment deadline after which users incur additional charges for delayed payment.

Article 46: In calculating the total amount payable by each user, 'The Operator' will consider the following:
- The meter reading for each user, detailing the monthly energy consumption. Users who for some special reason do not have a meter will have a special tariff that will be determined between 'The Proprietor' and the enterprise, during the period where energy cannot be metered. This period must under no circumstances be longer than six months.
- The prices established for electricity consumption, street lighting, delayed payments, interest and other services such as disconnection and reconnection.
- Taxes payable such as value added tax (VAT).

Heading X: Managing the income from the electricity services

Article 47: The following services generate income:
- Payment from the users for the electricity consumed.
- Payment for connection and reconnection of the service.
- Charges for delayed payment, interest or other payments relating to previous services.

A savings account will be opened jointly be the manager of 'The Operator' and the named manager of 'The Owner' in the case of municipalities. In other cases the holders will be defined in agreement with the participating members. The account will constitute the 'Electrification Funds'.

The income from the service will primarily be to pay for the services of the operator as agreed in the contract with 'The Proprietor'. 'The Operator' will pay themselves directly each month out of the income received.

The remaining balance, once 'The Operator's' fees and the operating costs in the contract have been paid, as well as the costs of maintenance and replacements, will be deposited by the manager of 'The Operator' in the savings account (Electrification Funds). She/he must prepare a statement of the income and expenditure for each month (after subtracting the charge for The Operator's services) a maximum of eight (8) working days after the payment deadline for the service.

Heading XI: Penalties for the operating enterprise

Article 48: Grounds for termination of contract are if 'The Operator':
- Fails to deliver, or repeatedly delays delivery of, the technical-commercial reports on the working of the electricity service.
- Shows repeated poor business management of the electricity service.
- Causes damage to the installations through deficient preventative maintenance.
- Uses the equipment with different ends to those stipulated in the contract. Grounds for termination of the contract are if 'The Owner':
- Repeatedly fails to pay for the provision of the service.
- Fails to address the need for equipment, tools, materials etc. necessary for the appropriate operation and maintenance of the electricity service, which were previously requested by the enterprise.

Heading XII: Final provisions

Article 49: Alterations to this Code of Conduct must be ratified in municipal by-laws.

Any citizen may present suggested alterations. These must be handed to 'The Owner' in writing. The proprietor must include them as agenda items in the next meeting, raising the consideration of the suggested alteration for debate and approval or filing.

Article 50: This Code of Conduct is an integral part of 'The Operator's' contract.

Date ....

Signed ....

CHAPTER SIX

Contract between the owner and the micro-enterprise operating and managing the electricity service

The document presents the principal parts that a contract for electricity services must include. It begins with the clear acknowledgement of the legal framework that forms a context for the contract and services.

Legal framework

The principal laws to consider in a contract of this type are: the Organic Law on Municipalities, the Law on Electricity Concessions in its articles about commercialization of electricity, and laws on small enterprise.

Purpose of the contract

The purpose of this contract must be clearly established, for example to set up an operating service for the energy supply scheme.

Contracting Party

This clearly defines the party, its location and its property over the scheme.

Part of the contract must contain an inventory (containing values) of all the infrastructure and additional equipment, materials and tools, donating them in the concession to the Contracted Party the strict purpose of performing the functions laid out in this contract.

Contracted Party

Trading name: ....
Address: ....
Name and electoral register number of the legal representative: ....
Type of enterprise: ....
RUC:[1] (where this applies) ....

Activities of the Contracted Party

The Contracted Party is committed to performing:
- Activities related to operation and maintenance:
  - operating and maintaining the structures, equipment and machines in good working order and according to an operation programme;
  - conducting preventative maintenance work on the equipment, installations and instruments, in accordance with an established preventative maintenance programme;
  - conducting civil works and preventative maintenance according to an established maintenance programme;
  - presenting quarterly reports on the maintenance work and suggesting completion dates for corrective maintenance activities.
- Activities related to the management of the electricity service:
  - operating the electricity networks and street lighting;
  - preventative maintenance of the networks;
  - installing spur lines to houses;
  - disconnecting and reconnecting the electricity supply, in accordance with the stipulations of the Code of Conduct for the operation of the electricity service;
  - meter reading;
  - customer service (complaints, user register etc.);
  - carrying out billing and registering income for the sale of energy;
  - invoicing and maintaining the register of users;
  - promoting the reasonable use of electrical energy among the users;
  - various coordination and control functions;
  - writing reports and monthly and annual statements on the working of the service.

Contract duration

The contract lasts for five (5) years. It is renewable if both parties agree and letters of acceptance have been exchanged. The five-year term is arbitrary, but it is a sensible length of time so that the enterprise can demonstrate its suitability and fulfil its role. The term can be longer; however, contract renewal is always a good measure.

Contractual sum

The basic sum of the contract is solXXX.YY (nuevos soles). This represents a fixed sum for the service of operation and maintenance of the installations.

Additionally the Contracted Party will receive a sum equivalent to 20 per cent of the amount invoiced monthly for the sale of electrical energy. This sum will go towards the management of the electricity service.

Method of payment

The Contracted Party will receive monthly payments for the fixed sum. This fixed sum will be paid on the fifth day after the payment deadline.

The payment deadline will be set each month by the Contracted Party according to the invoicing process.

Activities of the owner

Within the framework of this contract, the municipality is committed to:
- Make regular payments to the Contracted Party of the total, amounting from the sum of the fixed amount and the variable amount for the management of the electricity service.
- Provide the equipment (materials, parts, tools etc.) necessary for the preventative maintenance programme. This sum must be taken from the remaining balance once the Operator has received payment.
- Provide, in the form of a loan and for the duration of this contract, a premises for exclusive use as an office. If the contractual relation is terminated, the Contracted Party is committed to hand back the premises in the same state in which it was received and within no more than 10 days from severing the contract.

  Manage through a suitable organization the delivery of training courses in:
- Operating and maintaining the electromechanical equipment of the micro-hydro scheme.
- Operation and maintenance of the civil works of the micro-hydro scheme.
- Operation and maintenance of the distribution networks.
- Business management (administration, accounting, marketing etc.).
- Provide the Contracted Party with technical information about the equipment and other installations (manuals, plans etc.).
- Provide the Contracted Party with a Code of Conduct for the operation of the electricity service, which can be altered with mutual consent.

Causes for a contract to become null and void

The following situations will cause the contract to become null and void:
- Failure to deliver, or repeatedly deliver late, the technical-commercial reports on the working of the electricity service.
- Repeated deficiency in the commercial management of the electricity service. Some indicators of good efficiency are: few delayed payments, financial statements delivered punctually to the proprietor, punctual delivery of reports, rapid customer service, reliable work and reliable hours of operation according to the contract.
- Poor condition of the installations, caused by a deficient or absent preventative and/or corrective maintenance programme. These can occur owing to

negligent repair or maintenance work, inappropriate use of tools and/or services and other omissions, which hinder the smooth working of the scheme.
- Use of the equipment for different aims to those stipulated in the contract.

Additional clauses

The proprietor will not assume any responsibilities beyond those stipulated in the contract. The proprietor will not assume any responsibility for any accidents that may befall the staff of the Contracted Party. The Contracted Party must consider this issue, as the health of its operating staff will be its responsibility.

CHAPTER SEVEN

Electricity supply contract

This instrument is the contract for Electrical Energy Supply, which is signed by two parties: The first being THE ENTERPRISE OPERATING AND ADMINISTRATING THE ELECTRICITY SERVICE, represented by (name of proprietor), which will hitherto be known as THE ENTERPRISE, whose address is ...., represented by Mr...................., identified with identity card number .... The second party is Mr ...., with identity card number ...., residing at ...., who will hitherto be known as THE CLIENT, in the following terms and conditions:

First

Mr (name of owner) has contracted through a public tendering process the service of THE ENTERPRISE to operate and manage the generation, distribution and commercialization of energy as a Public Electricity Service in ....

Second

THE ENTERPRISE is committed to supplying electricity to the client in the form of alternating current of 220 volts .... in medium tension; single-phase .... or triple-phase .... in the premises located in .... No. ....

Third

The tariff option chosen by the client is established in THE ENTERPRISE's tariff system, the system of descending blocks, which applies in accordance with the agreements of the Auditing Committee and the proprietor.

Fourth

THE ENTERPRISE will take monthly readings for invoicing purposes. THE CLIENT must pay the amount corresponding to their consumption of electrical energy, street lighting, monthly fixed charge and charges for connection or reconnection, according to the tariff structure in effect, within fifteen (15) days of the date the invoice is sent out.

If the invoice is not paid within the period listed above, THE ENTERPRISE will apply a compensatory charge of X per cent a month on the sum outstanding, which constitutes a charge for delayed payment.

Fifth

THE ENTERPRISE will disconnect the service at a cost to THE CLIENT without having to give prior notice under the following circumstances:
- When payment has not been received for two invoices and/or charges which were duly notified.
- When electrical energy is consumed without the authorization of THE EN-TERPRISE or when the service conditions are broken.
- When the security of persons or properties are jeopardized through damage to the installations, whether these be under the management of THE EN-TERPRISE or be internal installations belonging to THE CLIENT (Art. 90 D.L. 25844).
- For maintenance work on the generating and/or distributing infrastructure.

Sixth

THE ENTERPRISE will reimburse THE CLIENT when a bill is inaccurate owing to the lack of an adequate meter reading or an invoicing error (Art. 92 D.L. 25844).

Seventh

THE ENTERPRISE can change the conditions of Supply in the case of force majeure. In this case it must give notice to THE CLIENT. THE ENTERPRISE is not responsible for the damage that the particular installations might suffer or for damage that such interruptions might cause.

Eighth

If THE CLIENT believes that the Public Electricity Service contracted is not performing in accordance with the stipulations of the Code of Conduct for Operation and Functions of the Electricity Service in (name of place)........, and/or aspects relating to the installation, invoicing, charging and others, he can present his complaint to THE ENTERPRISE, referring to the provisions in the Code of Conduct.

Ninth

This contract is governed by the Law on Electricity Concessions, its Code of Conduct, Rulings from the Commission of Electricity Tariffs, the Ministry of Energy and Mines and also by the Civil Code. Any aspect not referred to in this contract will be dealt with according to the legal provisions above, giving preference to the process of direct negotiation.

Tenth

This contract is effective for one year, starting from its signing. Its renovation is automatic unless either party intervenes, which must take place thirty (30) days before its end.

Eleventh

In the case that THE CLIENT transfers his rights over the property, the new owner will automatically stand in for him in the contract in all his rights and duties.

It is agreed between the two parties to submit the contracting parties to the jurisdiction of the judges and juries of .... They will submit to their ruling, renouncing other regional laws.

The two parties, in full knowledge of the contents of this document, accept it and proceed to sign it as a sign of their agreement, on the ...... day of the month of ...... in the year of ....

....

On behalf of the enterprise .... Client

Instruments applied in Conchán

In addition to the tariff structure described in Chapter 4, the following instruments were applied in the case of Conchán.

Code of Conduct for the operation and functions of the electricity service

Heading I: General orders

Article 1: In this Code of Conduct, when the terms 'The Law', 'The Code of Conduct', 'The municipality', 'The Operator', 'The Service' and 'The Committee' are used, they refer to the Law on Electricity Concessions, its Code of Conduct for the Operation and Functions of the Electricity Service, the District Municipality of Conchán, the Enterprise managing and operating the electricity service, the public electricity service in Conchán and the Auditing Committee, respectively.

Article 2: This Code of Conduct for the operation and functions of the electricity service in Conchán has the objective of regulating the activities related to the generation, distribution and commercialization of electrical energy in the district of Conchán, in the province of Chota, in the department of Cajamarca.

Article 3: In accordance with the stipulations of Article 7 of the Law No. 25844, the Law on Electricity Concessions, referring to the generation, transmission and distribution activities that do not require concession or authorization, as well as the relevant articles in the Law No. 23853 (the Organic Law on Municipalities), the District Municipality of Conchán has the power to manage and organize the public electricity service in the district of Conchán, applying the organizational approach most conducive to the community's interests.

Article 4: Having evaluated the repeated financial inconvenience caused to the municipality by an inappropriate organization of the Service and the different possible organizational alternatives, the municipality has decided to hand over the management and operation of the public electricity service to a local enterprise, as decreed in the Contract of Operation and Maintenance Services.

Article 5: The Public Electricity Service is offered by the operating enterprise, in representation of the municipality.

The Operator has a contract for a defined period of time to operate and manage the electricity services in Conchán for the municipality. Once this period is completed, the contract can be renewed according to the criteria and mechanisms in the contract.

Heading II: The electricity service

Article 6: The electricity service is defined as the right of the user to make use of the electrical energy within the limits and restrictions established in the Electricity Supply Contract, which the interested users must sign with 'THE OPERATOR' in representation of the municipality.

The electricity service is offered to a person or legal entity requesting the service (including for commercial uses), who acquires the status of User. The electricity service is offered at their premises.

Article 7: All the people or legal entities whose premises fall within the district limits may be considered as users if they meet the following criteria:
• being a legal adult;
• being able to prove that they reside in the premises as the owner or tenant.

Article 8: The electricity service can be offered to the users in three different types:
• single-phase electricity service at low tension;
• three-phase electricity service at low tension;
• three-phase electricity service at medium tension.

Article 9: The electricity service is offered through an electrical connection, the spur line. The spur line is defined as the electrical connection between the distribution lines and a point inside the user's premises.

The service provided by the operator refers to the supply of electricity from the point where the spur line begins. The maintenance of the spur line and the internal connections in the user's premises are the sole responsibility of the user, as is any occurrence caused by the state of repair of the spur line and the installations or inadequate manipulation of these.

Heading III: The functions and rights of the district municipality

Article 10: In accordance with the third article of this Code of Conduct, the municipality is responsible for organizing the electricity service. As such it will supervise its normal running and ensure that the Operator completes the activities it is committed to performing in the framework of the contract of services signed by both parties.

Article 11: The municipality will appoint a representative as a member of the Auditing Committee for the electricity service. The municipal representative's role will be defined by the other members of the committee, and their rights comply with the Code of Conduct for the functions of the Auditing Committee.

Article 12: Every month, the municipality will receive a financial statement and a monthly report on ordinary occurrences. It will be able to demand extraordinary reports from the Operator or Auditing Committee when the circumstances call for this.

Article 13: The municipality is responsible for paying the Operator the due amount at the times specified in the Service Contract signed by both parties.

Article 14: The municipality, under inventory, must provide the Operator with the materials, spare parts and tools necessary for the activities it is committed to perform. The lack of materials, parts and/or tools necessary for service delivery exonerates the Operator of responsibility in cases where its ability to perform its functions was hindered.

Article 15: As part of the concession, the municipality will provide the Operator with premises to be exclusively used as an office for the duration of the Contract of Service Provision, so that the Operator can conduct its activities to deliver the electricity service.

The conditions specific to the concession are regulated in the Contract of Service Provision signed by the Operator and the municipality.

Heading IV: The functions and powers of the operating and managing enterprise

Article 16: The Operator is responsible for:
- Producing electricity in accordance with user requirements in an efficient and uninterrupted manner, within the limitations of the machines and the electricity service scheme.
- The commercialization of the electrical energy produced, in accordance with cost-reducing criteria and in accordance with the tariff provisions established by the municipality.
- The maintenance of the equipment, the installations and the energy production and commercialization infrastructure in adequate working condition.
- The preparation of reports for the municipality and the Auditing Committee on the commercialization and production of electricity.

Article 17: In these activities, the Operator will take into account the following:
- The electricity must reach the users at the agreed voltage and frequency levels.

- Interruptions to the service must be attributable to unforeseeable circumstances and not to the negligence of its personnel.
- Customer service must be polite, cordial and timely.

Failure to comply with these obligations will incur the warnings and/or penalties defined by the municipality in accordance with current laws.

Article 18: The Operator is subject to the controls that the Auditing Committee exercises under ordinary and extraordinary circumstances.

Heading V: The supervision of the service

Article 19: The supervision of the electricity service activities is the responsibility of the 'Auditing Committee'. This committee is the highest authority created by the municipality to supervise, penalize and arbitrate in the case of disputes between any user on the Register of Users and the Operator in situations solely arising out of the provision of the electricity service.

Article 20: The Auditing Committee is made up of six (6) active members, who perform the following functions:
- president;
- vice-president;
- secretary;
- three members.

In this Committee, the users of the electricity service, the municipality and the Operator will be represented.

Article 21: The Committee's responsibilities and powers are:
- To adhere to and ensure the adherence to this Code of Conduct and approve any modifications and alterations to it, through the express request of any of the committee members.
- To ensure that the infrastructure for generating, commercializing and distributing electricity is maintained in good working order.
- To establish and apply the penalties for committee members fail to perform their duties.
- To name, appoint or dismiss the committee members, except the president, who is elected, appointed or dismissed in the communal assembly.
- To settle disputes between the users and the Operator, and penalise the party declared to be at fault.
- To ensure the tariff structure determined by the municipality is employed.
- To evaluate the monthly financial statements presented by the Operator and to offer opinions on its management with the aim of eventually renewing its contract.
- To hold meetings once a month.

Article 22: The responsibilities of the president are:
- To act as the official representative of the Committee.
- To adhere to and ensure adherence to this Code of Conduct and uphold the Committee's rulings.
- To ensure that the other committee members perform their activities.
- To call and chair Auditing Committee meetings (regular and extraordinary).
  In the absence of the president, the vice-president will take their place, assuming all the prerogatives and responsibilities of the president.

Article 23: The responsibilities of the secretary are:
- To keep an up-to-date record of the minutes, rulings and files of the Committee.
- To handle the internal and external correspondence and sign documents with the president.
- To keep other members informed of agreements and rulings of the Committee, keep attendance lists and keep note of the quorum, noting in the minutes the whole text of rulings and initiating the corresponding procedures.
- To call the meetings of the Auditing Committee, summoning third parties in the cases where this is required.

Article 24: The responsibilities of the members are:
- To participate actively in the regular and extraordinary Committee meetings.
- To participate in the commissions formed to investigate specific cases.
- To gather and channel the users' complaints.
- To disseminate to users the contents of this Code of Conduct and any alterations to it.

Article 25: The organization of the Committee is the responsibility of the active members in the regular and extraordinary meetings. The Committee can appoint special commissions in specific cases, with clearly defined objectives and for defined periods of time.

Article 26: The regular meeting is the highest authority of the Committee. It is made up of all the active members, with equal rights as full members.

Article 27: Regular meetings of the Committee will be conducted within five (5) days of receiving the monthly report from the Operator and for each member. Attendance at the meeting is absolutely compulsory.

Article 28: Extraordinary meetings will only be held to resolve important issues, which were not foreseen in the regular meetings. The extraordinary meeting will take place when required by the president of the Committee or when half plus one of the total members of the Committee request it.

Article 29: The quorum for regular and extraordinary committee meetings is half plus one of the total of the members. If the minimum for the quorum is not present, the president of the Committee will postpone the meeting by fifteen (15) minutes. After this time, the session will have to be completed with the members present, and the absences will be noted in the minutes.

Article 30: The members of the Committee do not receive payment, since no material advantage is pursued. Active members of the Committee will be entitled to a 50 per cent discount on their respective monthly electricity bills. The members of the Committee who are stripped of their position will automatically lose the benefit of the 50 per cent discount.

Article 31: Criteria for membership of the Committee are as follows:
• Being a legal adult.
• Not holding a public position that by its nature compromises the impartiality of the person.
• In the case of inhabitants from other areas, at least five (5) years' continuous permanent residence in the area where the service is provided.

Article 32: The members of the Committee will hold their positions for a period no longer than one year. They can renew their memberships if they receive a vote of confidence for outstanding performance in a communal assembly.

Article 33: It is the responsibility of the users to pay the monthly bills received from the Operator. If the user disagrees with the total payable or one of the figures on the bill, they must present their verbal complaint to the Operator, who must deal with the complaint immediately. If the Operator does not uphold the complaint, the user is free to appeal to one of the members of the Committee. The member can collect the complaint and can also request a report on the case from the Operator.

The Committee member who received the complaint, and who knows the position of both parties, will present it in the regular Committee meeting, adding a personal opinion. After checking against this Code of Conduct, the contract and other relevant documents and, in keeping with previous rulings, the Committee members will discuss the case exhaustively and then proceed to vote on it.

Once the result of the vote has been made public, the case will be considered closed. The result of the vote will be recorded in the minutes, as well as a brief report, which must be presented to the parties involved.

Article 34: The decisions or rulings of the Committee are final. However, the party who consider themselves affected by the Committee's decision are free to take their case to the judiciary courts of Chota.

Article 35: The members' discussions within the Committee must take place in a spirit of order, initiative, goodwill, impartiality and discipline, referring to this Code of Conduct as a framework.

Within the Committee, it is absolutely prohibited to defend or promote any political, religious, racial or social ideology. To do so repeatedly is considered a grave fault in members.

Heading VI: Tariffs

Article 36: The prices for the sale of electrical energy, the various charges (for street lighting, disconnecting or reconnecting electricity supplies, arrears on delayed payments, connection costs etc.) and the tariff for financial penalties for users (vested interest, resale of energy, clandestine use etc.) are established annually and take effect from the first of January of each year.

Article 37: The tariff is adjusted to the specifications established according to the system of descending blocks. The entire population must be made aware of this through a poster hung on the wall inside the Operator's office.

The results of the tariffs applied must be evaluated annually, taking into consideration the following:
• Projected annual average operation and maintenance costs for the generation, distribution and commercialization of electrical energy. The cost of materials, parts, tools and the payment to the operating enterprise must be taken into account.
• Generation capacity of the hydro-electric scheme in kWh per year.
• Projected income from the sale of energy.

Article 38: The application of the tariff is the responsibility of the Operator, while the Committee is in charge of supervising its correct application.

Heading VII: Collecting payment, delayed payments and customer service

Article 39: Collecting payment is an activity corresponding to the commercialization of electrical energy. Therefore it is the responsibility of the Operator and is subject to the supervision of the Auditing Committee.

The Operator will inform the users of the payment deadline with at least fifteen (15) days' notice, using the most effective communication medium.

It is the responsibility of the users to find out the deadline for payment. Therefore not knowing the deadline for payment will not be considered a valid reason for delayed payment for the energy consumed in a month.

Article 40: The deadline for payment determines the date until which the user can pay for the energy consumed without incurring additional charges.

After the deadline, the user will have to pay an additional charge for defaulted or delayed payment. Each year, the municipality will set the rate for each additional day. This charge for late payment has the aim of:

• Covering the administrative costs generated by the delay in payment, since staff have to commit time to this activity instead of another service activity.
• Creating awareness amongst the users of the negative effect of delayed payment on the normal working of the electricity service.

Article 41: Opening hours and working days for receiving payment for energy consumed, complaints, reports or queries must be communicated by the Operator using the most appropriate media.

Article 42: Customer service is a priority activity for the Operator. Politeness, efficiency and timeliness of service to users are indicators of the Operator's management of the electricity service. These will be considered by the municipality and the Committee when renewing the Operator's service contract.

Heading VIII: Disconnecting and reconnecting users

Article 43: When a specific user has failed to pay three (3) months' worth of energy consumed, their electricity service will be disconnected.

This service will only be reconnected by the Operator once the user has paid the debt and the additional charges for delayed payment, interest and administrative charges.

Article 44: The disconnection and reconnection of the service incur a cost, which must be met by the user as stipulated in the previous article.

The costs of disconnection and reconnection incurred through the maintenance, renovation or improvements to the infrastructure for distributing electrical energy will be met by the municipality.

Heading IX: Invoicing and the register of users

Article 45: The administrational activity of invoicing is performed by the Operator every month, to calculate the total amount payable by each user for the amount of energy consumed, street lighting etc. and to send out the invoices and prepare statistical information. Invoicing includes the printing of the invoice for each user, the generation of statistical information about sales and production of electricity (in financial terms and in units of energy) and the payment deadline after which users incur additional charges for delayed payment.

Article 46: In calculating the total amount payable by each user, the Operator will consider the following:

• The meter reading for each user, detailing the monthly energy consumption. Users who for some special reason do not have a meter will have a special tariff that will be determined by the municipality, during the period where energy cannot be metered.

- The prices established by the municipality for electricity consumption, street lighting, delayed payments, interest and other services such as disconnection and reconnection etc.
- Taxes payable such as value added tax (VAT).

Heading X: Penalties for the operating enterprise

Article 47: The contract can be terminated if the managing enterprise:

- Fails to deliver or repeatedly delays delivery of the technical-commercial reports on the working of the electricity service.
- Shows repeated poor business management of the electricity service.
- Causes damage to the installations through deficient preventative maintenance.
- Uses the equipment in the contract with different ends to those stipulated in the contract.

  Grounds for termination of the contract on the municipality's part are if it:
- Repeatedly fails to pay for the provision of the service.
- Fails to address the need for equipment, tools, materials etc. necessary for the appropriate operation and maintenance of the electricity service, which were previously requested by the enterprise.

Heading XI: Final orders

Article 48: Alterations to this Code of Conduct can only be made by the communal assembly.

Any citizen may present suggested alterations. These must be handed to the municipality in writing. The municipality must include them as agenda items in the next meeting, raising the consideration of the suggested alteration for debate and approval or filing.

Conchán, August 1999.

Contract for operating and managing the electricity service of Conchán

This document is a CONTRACT FOR OPERATING AND MANAGING THE ELECTRICITY SERVICE IN CONCHÁN. It is signed by the CONTRACTING PARTY: the DISTRICT MUNICIPALITY OF CONCHÁN, represented by the Mayor, Professor Diego Andrés Guevara Tarrillo, identity card No. 27369943, whose address is Jr. Grau No. 198 in the district of Conchán, in the province of Chota, in the department of Cajamarca. The other party signing is the SERVICE PROVIDER: the SAN ISIDRO ELECTRICITY SERVICES ENTERPRISE OF CONCHÁN, publicly registered with number 166-Chota-28/01/99, whose address is Jr. Grau s/n Conchán and is represented by the General Manager,

Mr Wilder Alvarado Pita, identity card No. 27381100; under the following terms and conditions:

First: The CONTRACTING PARTY is the proprietor of the infrastructure for generating, distributing and commercializing electricity (an inventory of all the goods that make up this infrastructure is part of the present contract as ANNEX 1) installed in this area and currently operational. As such, given the need to contract the services of an operator and manager of this electricity service infrastructure, it CONCEDES the contract to the SERVICE PROVIDER, winner of a public tendering exercise (the details of which are part of this contract as ANNEX 2) to operate and manage the infrastructure. Through this operation and management, it must provide electricity services for the people of the district of Conchán. To this effect, the CONTRACTING PARTY concedes the infrastructure detailed above to the SERVICE PROVIDER.

Second: For its part, the SERVICE PROVIDER receives the infrastructure conceded, according to a concession document, and commits itself to performing:

- Activities related to operation and maintenance:
  - Operating and maintaining the structures, equipment and machines in good working order and according to an operation programme.
  - Conducting preventative maintenance work on the equipment, installations and instruments, in accordance with an established preventative maintenance programme with this aim.
  - Conducting civil works and preventative maintenance according to an established maintenance programme with this aim.
  - Presenting quarterly reports on the maintenance work, suggesting completion dates for corrective maintenance activities and the necessary equipment, materials, tools and parts required.
- Activities related to the management of the electricity service:
  - Operating the electricity networks and street lighting.
  - Installing spur lines in premises.
  - Disconnecting and reconnecting the electricity supply, in accordance with the stipulations of the Code of Conduct for the operation of the electricity service.
  - Meter reading.
  - Customer service (queries, complaints, user register etc.).
  - Invoicing and maintaining the register of users.
  - Carrying out billing and registering income for the sale of energy.
  - Promoting the reasonable use of electrical energy among the users.
  - Various coordination and control functions.
  - Writing reports, monthly and annual statements on the working of the service to the CONTRACTING PARTY, in the format required by them.
  - Strict observance for the staff, with whom the CONTRACTING PARTY has no contractual link.

Third: In return, the SERVICE PROVIDER will receive from the CONTRACTING PARTY a fixed monthly sum of up to 1100 /100 Nuevos Soles, the same amount awarded for the operation and maintenance of the installations.

In addition, the SERVICE PROVIDER receives an allowance equivalent to TWENTY PER CENT (20%) of the income collected that month from the sale of electrical energy. This is on top of the fixed amount described above and after an amount has been set aside for maintenance and repair or replacements. This goes towards administrative and functional costs met by the SERVICE PROVIDER.

The fixed sum will be paid on the fifth (5) day after the payment deadline for users, as will the percentage payment.

The payment deadline will be 15 days after the invoice has been sent out on a monthly basis by the SERVICE PROVIDER, according to the invoicing process.

Fourth: For its part, the CONTRACTING PARTY is committed to:

- Make regular payments to the SERVICE PROVIDER, of the total amounting from the sum of the fixed amount and the variable amount for the management of the electricity service, setting aside an amount for maintenance and repair or replacements.
- Provide the equipment (materials, parts, tools etc.) necessary for the preventative maintenance programme.
- Provide, as part of the concession and for the duration of this contract, a premises for exclusive use as an office. If the contractual relation is terminated, the SERVICE PROVIDER is committed to hand back the premises in the same state in which it was received and within no more than 10 days from severing the contract.
- Manage through a suitable organization the delivery of training courses in:
  - Operation and maintenance of the electromechanical equipment of the micro-hydro scheme.
  - Operation and maintenance of the civil works of the micro-hydro scheme.
  - Operation and maintenance of the distribution networks.
  - Business management (administration, accounting, marketing etc.).
- Provide the SERVICE PROVIDER with technical information about the equipment and other installations (manuals, plans etc.).
- Provide the SERVICE PROVIDER with a Code of Conduct for the operation of the electricity service, which can be altered by mutual consent.
- Coordinate with the SERVICE PROVIDER the performance of activities required by laws regulating the generation, distribution and commercialization where applicable.

Fifth: This contract will last for a period of five (5) years, renewable by mutual agreement between the two parties if letters of acceptance have previously been exchanged.

Sixth: This contract will be terminated if general commercial or civil laws are infringed or in the following situations caused by the SERVICE PROVIDER:

- Failure to deliver, or repeatedly late delivery of, the technical-commercial reports on the working of the electricity service.
- Repeated deficiency in the commercial management of the electricity service. Some indicators of good efficiency are: few delayed payments, financial statements delivered punctually to the proprietor, punctual delivery of re-

ports, rapid customer service, reliable work and hours of operation according to the contract.

- Poor condition of the installations, caused by a deficient or lacking preventative and/or corrective maintenance programme. These can occur owing to negligent repair or maintenance work, inappropriate use of tools and/or services and other omissions, which hinder the smooth working of the scheme.
- Use of the equipment in the contract for different aims to those stipulated in the contract.
- Other situations that imply a failure to comply with the aims of the contractual relationship.

Additional clauses

1 The owner will not assume any responsibilities beyond those stipulated in this contract. Within this context, the proprietor will not assume any responsibility for any accidents that may befall the staff of the contracted party. The CONTRACTING PARTY only has a contractual relationship of service provision with the SERVICE PROVIDER and not with individuals.

2 However, in accidental cases (electrical discharges, blockages to the water flow etc.) THE ENTERPRISE will call an emergency meeting with the CONTRACTING PARTY to provide a timely solution and defined responsibilities.

3 The total payment to the SERVICE PROVIDER, which is the sum of the fixed amount plus 20 per cent of the monthly income collected, will be reviewed annually, considering the economic results of the management of the service. This amount for managing the service must never be more than the monthly income from the sale of electricity.

With the agreement of both parties to all the stipulations, we sign this document in the area of Conchán, on the .... day of the .... month of the year 2000.

....

....

DIEGO ANDRES GUEVARA TARRILLO WILDER ALVARADO PITA
L.E. No. 27369943   L.E. No. 27381100
ON BEHALF OF THE   ON BEHALF OF THE SERVICE
CONTRACTING PARTY   PROVIDER

....

TECHNICAL ADVISER, Practical Action – LA

Electricity supply contract

This document is the electricity supply contract, which is signed by two parties: the first being THE ENTERPRISE OPERATING AND ADMINISTRATING THE

ELECTRICITY SERVICE, representing the district municipality of Conchán, which will hitherto be called THE ENTERPRISE, whose address is Jr. Grau s/n in this city, duly represented by Mr Wilder Alvarado Pita, whose identity card number is No. 27381100. The second party is Mr .... with identity card number .... residing at ...., who will hitherto be known as THE CLIENT, in the following terms and conditions:

First

THE MUNICIPALITY has contracted through a public tendering process the service of THE ENTERPRISE to represent it, in operating and managing the generation, distribution and commercialization of energy as a Public Electricity Service in the district of Conchán.

Second

THE ENTERPRISE is committed to supplying electricity to the client in the form of alternating current of 220 volts .... in medium tension; single-phase .... or triple-phase .... in the premises located in .... No. ....

Third

The tariff option chosen by the client is the BT-5 .... BT-6 .... or another. The tariff will apply for the period of one year (this option can be altered on receipt of a written request from THE CLIENT, but otherwise applies annually).

Fourth

THE ENTERPRISE will take monthly readings for invoicing purposes. THE CLIENT must pay the amount corresponding to their consumption of electrical energy, street lighting, monthly fixed charge and charges for connection or reconnection, according to the tariff structure in effect, within fifteen (15) days of the date the invoice is sent out.

If the invoice is not paid within the period listed above, THE ENTERPRISE will apply a compensatory charge of 2 per cent a month on the sum outstanding. This constitutes a charge for delayed payment.

Fifth

THE ENTERPRISE will disconnect the service at a cost to THE CLIENT without having to give prior notice under the following circumstances:
- When payment has not been received for two invoices and/or charges which were duly notified.

- When electrical energy is consumed without the authorization of THE EN-TERPRISE or when the service conditions are broken.
- When the security of persons or properties are jeopardized through damage to the installations, whether these be under the management of THE EN-TERPRISE or be internal installations belonging to THE CLIENT (Art. 90 D.L. 25844).
- For maintenance work on the generating and/or distributing infrastructure.

Sixth

THE ENTERPRISE will reimburse THE CLIENT when a bill is inaccurate due to the lack of an adequate meter reading or an invoicing error (Art. 92 D.L. 25844).

Seventh

THE ENTERPRISE can change the conditions of Supply in the case of force majeure. In this case it must give notice to THE CLIENT. THE ENTERPRISE is not responsible for the damage that the particular installations might suffer or for damage that such interruptions might cause.

Eighth

If THE CLIENT believes that the Public Electricity Service contracted is not performing in accordance with the stipulations of the Code of Conduct for Operation and Functions of the Electricity Service in Conchán, and/or aspects relating to installation, invoicing, billing and others, he can present his complaint to THE ENTERPRISE, referring to the provisions in the Directoral Resolution No. 012/95-EM/DGE.

Ninth

This contract is governed by the Law on Electricity Concessions, its Code of Conduct, Rulings from the Commission of Electricity Tariffs, the Ministry of Energy and Mines and also by the Civil Code. Any aspect not referred to in this contract will be dealt with according to the legal provisions above, giving preference to the process of direct negotiation.

Tenth

This contract is effective for one year, starting from its signing. Its renovation is automatic unless either party intervenes, which must take place thirty (30) days before its end.

Eleventh

If THE CLIENT transfers his rights over the premises, the new owner will automatically stand in for him in the contract in all his rights and duties.

It is agreed between the two parties to submit the contracting parties to the jurisdiction of the judges and juries of Chota. They will submit to their ruling, renouncing the laws of their local area.

The two parties, in full knowledge of the contents of this document, accept it and proceed to sign it as a sign of their agreement, on the ...... day of the month of ...... in the year of ....

....
On behalf of the enterprise   Client

ANNEX

Rules for the public tendering process to select a company to manage the electricity service in Conchán:

Contents

1) Specific rules
    1.1   The contracting body
    1.2   Objective
    1.3   Funding
    1.4   Type of selection process
    1.5   Estimated budget
    1.6   Labelling the tenders
    1.7   Duration of service for the Operating Enterprise
2) General rules
    2.1   Purpose of the rules
    2.2   Legal basis
    2.3   Technical proceedings
    2.4   Companies that can participate
    2.5   Invitation to participate
    2.6   Queries
    2.7   Receipt of applications
    2.8   Opening the envelopes
    2.9   Awarding the contract
        2.9.1   Scoring on price
        2.9.2   Scoring on business experience
        2.9.3   Scoring on qualified management staff
        2.9.4   Scoring on qualified technical staff
        2.9.5   Summary of evaluation and grading
    2.10 Cases in which the selection process can be declared invalid

1) Specific rules

1.1 The contracting body

The district municipality of Conchán, whose address is Jr. Grau No. 198, in the province of Chota, department of Cajamarca.

1.2 Objective

The district municipality of Conchán invites this public tendering process with the objective of selecting those enterprises that propose the most appropriate technical and economic solutions for providing services in 'the management of the local electricity service'.

1.3 Funding

The service that the winning operating enterprise will provide will be covered by its own resources generated through the sale of electricity to the users.

1.4 Type of selection process

This public tendering process will be a CLOSED process, without preselection and without funding, without reimbursement costs and at high cost. It will only include the applicants, enterprises or individuals that are from the district of Conchán at the time of the selection process.

1.5 Estimated budget

The estimated budget for the provision of the management service for the current selection process is S/.............. Nuevos Soles.

This figure will be reviewed annually, taking into account the economic results of the management of the service. Under no circumstances will the figure for the management service exceed the monthly income from the sale of electricity.

1.6 Labelling the tenders

The envelopes containing the applicants' tenders will be labelled as follows:

CONCHÁN DISTRICT MUNICIPALITY

Evaluation Committee

PUBLIC TENDER No.: ....

RULES FOR THE PUBLIC TENDERING PROCESS TO SELECT AN ENTERPRISE TO MANAGE THE ELECTRICITY SERVICE IN CONCHÁN

Representative for the applicant:

Mr ....

Address: ....

1.7 Duration of service of the Operating Enterprise

The managing enterprise will be responsible for the management of the service for a period of five (5) consecutive years (the period stipulated in the contract). This period can be extended for a similar period following open negotiations between the enterprise and the contracting municipality.

If the municipality and the managing enterprise are not able to renew the contract of service, the municipality will be required to call a new public tendering process.

2) General rules

2.1 Purpose of the rules

The purpose of these rules is to establish the norms and procedures that the competing enterprises will undergo to provide the management of the local electricity service.

2.2 Legal basis

This public tendering process is governed by the following rules:
• organic Law on Municipalities No. 23853;
• the Law on State Contracting and Acquisitions, No. 26850;
• transitory orders on the application of Law No. 26850;
• sole text on Administrative Procedures of the District Municipality;
• the Technical Proceedings.

2.3 Technical Proceedings

The Technical Proceedings of the competition are made up of:
• specific rules;
• general rules;
• evaluation system;
• annexes;
• terms of reference of the contract.

2.4 Companies that can participate

The tendering process is 'open' without any kind of restrictions, to all enterprises or individuals that are interested in managing the local electricity service and believe themselves to meet the minimum criteria.

In the case where one or more people, without official legal recognition as an enterprise, want to participate, they can do so as long as they include with their application a letter of intent to formalize their legal status if they are awarded the contract.

2.5 Invitation to participate

The invitation to participate will be issued through communal assemblies taking place between the 2nd and 5th of September 1998, at the times advertised in the municipal premises.

All interested parties can participate in these assemblies, if they can demonstrate that they live in the area and have completed at least their primary education.

2.6 Queries

Applicants' queries or observations on this document can be addressed verbally and in full in any of the assemblies between the 3rd and 5th of September of this year.

The queries will be resolved by the members of the EVALUATION COMMITTEE formed to oversee this tendering process.

2.7 Receipt of applications

Applications will be formally received between the 4th and 6th of September of this year.

The applications will be received in the municipal premises during office hours.

They are to be submitted in a sealed envelope and will be kept safe until the opening date to be evaluated and graded. Under no circumstances should the envelopes be opened before the opening date. If this happens, the process is open to challenge.

Any applicants presenting altered, fraudulent or unsuitable documents, whenever this is detected, will be disqualified, even if they were awarded the contract. They will be prohibited from participating in future tendering processes or bidding for future work.

The applications must contain:

- An economic estimate for the provision of the service.
- The names of the management staff, specifying names, surnames, ages and level of education.
- The names of the technical staff, specifying names, surnames, ages and level of education.
- One sheet on the work experience of each of the management and technical team members.
- A letter of intent to formalize the status of the enterprise, if they do not have legal status.

If any of these documents is missing, the applicant will be disqualified.

2.8 Opening the envelopes

The opening of the envelopes will take place on the 5th of September of this year.

The EVALUATION COMMITTEE will document the sequence of events from the opening of the envelopes to the evaluation and grading of the tenders.

Once the process of evaluation and grading has been finalized, the EVALUATION COMMITTEE will publish the scores that the different applicants achieved in a league table.

2.9 Awarding the contract

The contract will be awarded to the tender that achieves the highest score. The scores will be published in the municipal premises on the 8th of September of this year.

Scores will be awarded according to the following criteria:

2.9.1 Scoring on price

The score is the result of dividing the Estimated Budget by the sum requested by the applicant under number 1.5, multiplied by 10.

$$\text{Price score} = \frac{\text{Estimated budget}}{\text{Sum requested}} \times 10$$

The figure calculated will be rounded up to two decimal points.

2.9.2 Scoring on business experience

No experience	0 years	2 points
Minimal level of experience	0 to 5 years	5 points
Medium level of experience	5 to 10 years	10 points
High level of experience	10 or more years	15 points

2.9.3 Scoring on qualified management staff

No formal qualifications	5 points
Technical training	10 points
High level of management training	15 points

2.9.4 Scoring on qualified technical staff

No formal qualifications	5 points
Technical training	10 points
High level of technical training	15 points

2.9.5 Summary of evaluation and grading

No.	Item:	Maximum possible score:
1	Scoring on points	Variable
2	Scoring on business experience	15
3	Scoring on qualified management staff	20
4	Scoring on qualified technical staff	20

2.10 Cases in which the selection process can be declared invalid

The selection process can be declared invalid if at any stage of the process there are not two (2) or more applicants.

If the selection process is declared invalid, there will be no reimbursement for the participating applicants.

2.11 Evaluation of applications

The EVALUATION COMMITTEE will evaluate and grade the documentation presented. The committee is made up of two members of the municipality, the Mayor and a municipal councillor, two technical advisers from Practical Action and two representatives of the users.

The members will choose one member to preside over the committee, through the mechanism they see fit.

2.12 Dealing with complaints and challenges

Complaints or challenges will be presented in writing to the EVALUATION COMMITTEE until the 6th of September of this year.

Only the individuals or representatives who participated in the process have the right to challenge the process.

The EVALUATION COMMITTEE will be responsible to evaluate the challenges and rule in favour or against them by the 6th of September 1998 at the latest.

2.13 Duties of the applicant awarded the contract

If the applicant awarded the contract does not have official legal status, they are committed to apply for it within 15 days of the results of the process being published.

Legal status as an enterprise is an indispensable requirement for the contract of services to be signed.

The winning enterprise will appoint a representative to carry out an inventory of equipment and installations that make up the infrastructure of the electricity service, to determine the state in which the installations will be received.

The winning enterprise, and all the staff in its staff listing, are committed to participate actively in the training days. A specialist organization will be hired to provide this training as a measure to strengthen the enterprise.

3) Evaluation system

This section outlines the methodology used to evaluate and grade the applications and how the league table is determined.

3.1 Scoring on price

The scoring on price is the score achieved according to the sum requested as given in Section 2.9.1.

3.2 Scoring on business experience

The scoring on business experience is the score that the applicants will achieve according to the extent of work experience of each of the people in the staff list.

Business experience will be measured in years of work for each individual. For the grading, the most experienced individual will be considered. Applicants only need to present a table summarizing the names and surnames of each person, years working in a business and the position held.

The organization awarded the contract will have to produce relevant documents as evidence of the years of experience of the person taken, as a reference, within 10 days of the league table being published. Failure to present these documents will disqualify the organization, and the second applicant in the league table will be accepted as the winner.

3.3 Scoring on qualified management staff

The scoring for qualified management staff is the score received by the applicants according to the highest education level of the individuals mentioned in the staff list who will be performing management functions.

The scores for management staff will be in terms of the highest educational level obtained by each individual. For the grading, the individual with the highest educational level will be considered. Applicants only need to present a summary table with the names and surnames of each individual, the level of education and the institution where the individual studied.

The applicant awarded the contract will have to produce relevant documents as evidence of the educational level of the person taken, as a reference, within 10 days of the league table being published. Failure to present these documents will disqualify the organization, and the second applicant in the league table will be accepted as the winner.

3.4 Scoring for qualified technical staff

The scoring for qualified technical staff is the score received by the applicants according to the highest education level of all the individuals mentioned in the staff list who will be performing technical functions.

The scores for technical staff will be in terms of the highest educational level obtained by each individual. For the grading, the individual with the highest technical educational level will be considered. Applicants only need to present a summary table with the names and surnames of each individual, the level of education and the institution where the individual studied.

The applicant awarded the contract will have to produce relevant documents as evidence of the technical level of the person taken, as a reference, within 10 days of the league table being published. Failure to present these documents will disqualify the applicant, and the second tender in the league table will be accepted as the winner.

4) Additional orders

4.1 Role of the Evaluation Committee

4.1.1 Functions
- To evaluate the tenders presented by the enterprises.
- To apply the correct procedures and determine a winner.
- To declare the selection process invalid if there are not enough applicants or the applicants fail to meet the criteria for the service concession.
- To inform the users on the selection process carried out.
- To document with the District Council the results of the evaluation.
- To resolve any obstacles, misunderstandings or disputes resulting from this document at the request of any of the applicants.

4.1.2 Responsibilities
- Inviting local participation in the selection process.
- Determining the period of invitation, evaluation and selection of the winner.
- Actively participating in the evaluation and selection process.
- Being as transparent as possible in the evaluation process.
- Resolving queries about the issues.
- Presenting a summary document on the process.

4.1.3 On the duration of the Evaluation Committee
- The Evaluation Committee only exists for the duration of the selection process for the enterprise or individual who wins the contract, including the signing of the service contract.
- If the process is postponed, the Committee will meet again to establish the schedule for a new process.
- Once the selection process is complete, there is a documented winner and the contract has been signed, the committee is permanently dissolved.

4.2 Representative of the organization applying

The tender must be signed by the representative of the organization applying. S/he will be the legal representative of the enterprise formed as a result of winning this tender process.

4.3 Using this information

The applicant may only use this information in this selection process. The municipality reserves the rights to take legal action in the case of any of the applicants making use of this information for other purposes.

4.4 If the winner fails to complete the process

If the winner awarded the contract will not continue with the process, their decision must be sent in writing to the EVALUATION COMMITTEE or the

Municipal Council (if the Committee has been dissolved) within five days of the publication of the league table. The Committee can appoint the applicant who was in second place in the league table as the winner, or to declare the selection process invalid.

4.5 Support for the winner of the selection process

- The municipality will loan the winner a premises to establish an office.
- The training offered by the municipality to strengthen the organization will take place once the contract has been signed and before the enterprise begins its operations.
- To invoice, maintain the register of users and write reports etc., the winning enterprise will be loaned a computer, equipped with a programme that supports these activities.

5) Annexes

5.1 Technical specifications of the service

5.1.1 Civil works

a) Intake: The intake is located at the source of the Concháno river. Before the micro scheme was built, there was a canal dating from over 50 years ago. The intake consisted of a few stones and branches to direct the water towards this canal. At present the intake is partially built of concrete. It is not appropriate to build a large structure because while the bases were built the source could change its position a few metres down.

b) Canal: The canal is trapezoidal and the last 18 metres are lined throughout. Near the settling basin it is rectangular and 300 metres long, with a marked slope all the way across since the widening and lining of the canal followed the course of the old canal. The last 18 metres leading to the forebay have an adequate slope that makes it easy to clean.

c) Settling basin–forebay: As a structure, the settling basin and forebay are in good condition. It consists of a gate before the canal meets the forebay, a rack. This controls the entry of water to the joint forebay–settling basin. In combination with the salt sluice gate, they make a perfect complement for the cleaning of the settling basin.

d) Penstock: The penstock is made of steel, with a 15-inch-diameter. It has three anchors, one at the beginning in the forebay, another where the section gradient changes and another at the end arriving at the power house. Each has its respective expansion joint.

e) Power house: The power house is constructed of concrete and bricks. Its roof is made of galvanized corrugated sheets. Inside the power house is the electromechanical equipment: the turbine, alternator, switchboard and controller. In a separate part is the low–high transformer.

f) Spillway channel: The spillway channel evacuates the water from the turbine towards the intake of the Concháno Tunnel. It is in good working condition.

5.1.2 Electromechanical equipment

a) Turbine

Brand	Hidrowerke S.A. (Peru)
Type	Francis
Power	80 kW
Net head	37.5 m
Discharge	0.3 m3/s
Speed	12000 RPM
Type of valve	12-inch gate

The turbine has a pressure gauge and vacuum gauge to measure the pressure and vacuum respectively.

b) Controller: The controller is of the automatic speed type. It has not functioned automatically since its installation, only manually. This requires an operator to review the voltage changes 24 hours a day, and to increase or decrease flow manually.

It currently has an electronic load controller with the following characteristics:

c) Electricity generator

Brand	GCZ Ingenieros
Model	G656
Series No.	940002
Voltage	400/231
Amperage	262.4 A
Power	100 kVA
Frequency	60 Hz
Speed	1200 RPM
Phases	3
Connection	Star connection
Insulation	F
Cos θ	0.8
Brushes	None
Manufacture standard	B5 4999/B5 5000
Servicing requirements	Continuous

d) Switchboard

The switchboard is a single item, installed near the generator. It contains the following measuring instruments:

- 1 ammeter, calibrated from 0 to 600 A, with an ammeter change-over switch for charge phases R, S and T.
- 1 voltmeter, calibrated from 0 to 500 V, with a voltmeter change-over switch for RS, ST, TR.
- 1 frequency meter, calibrated from 55 to 65 Hz.
- 1 kilowattmeter, calibrated from 0 to 100 kW.
- 1 energy meter, to measure energy used.

5.1.3 Electricity networks

a) Primary network: The primary network is the high tension network that includes the step-up transformer, the transmission line, the step-down transformer and the switchboard.

b) Step-up transformer

Brand	Elko Peruana S.A.	
Power	100 kW	
Number of phases	3	
Frequency	60 Hz	
Series No.	2555150	
Standard	370002	
Cooling	ONAN	
Weight of oil	231 kg	
Height	3000 m above sea level	
Total weight	685 kg	
Year of manufacture	1994	
Tension	10,000 V	400–231 V
Peak current	5.773 A	144.3–249.9 A
Insulation level	28 kV	2.5 kV
Connection	Y	

c) Transmission line: The transmission line is made up of three conductors of 10 mm², which conduct a voltage of 10,000 volts in a conduit of 2 km from the power house to the town, supported by 11-m-long posts.

d) Step-down transformer: This transformer is in the elevated substation in a road in the town. It is supported by two concrete posts of 12 m length. Its aim is to transform the high voltage of 10,000 V to the low tension, in this case 400–231 V. The specifications of the transformer are as follows:

Brand	Elko Peruana S.A.
Power	100 kW
Number of phases	3

Frequency	60 Hz	
Standard	370002	
Cooling	ONAN	
Weight of oil	231 kg	
Height	3000 m above sea level	
Total weight	685 kg	
Year of manufacture	1994	
Tension	400–231 V	10,000 V
Peak current	144.3–249.9 A	5.773 A
Level of isolation	2.5 kV	28 kV
Connection	Y	9
Terminals	6	3

e) Switchboard: The switchboard is located in the town's substation, with the step-down transformer. The three distribution circuits for the population are from this switchboard, both for household electricity and for street lighting. The connection for the street lighting circuit is manual.

f) Secondary network and household spur lines: The secondary distribution network involves three circuits that run across the streets of the town to reach each of the houses. There are also three circuits for street lighting, reaching each of the lights installed on the town roads.

The same distribution system for the secondary network is three-phase 380–220 V tension, three-phase with 380 V tension between phases and a neutral for particular service and single-phase for the street lighting with the common neutral, obtaining a tension of 220 V.

5.1.4 General characteristics of the service
No. of users
No. of potential users
No. of users with a meter
No. of bills
Delays in bill payment

5.2 Template for letter formalizing the enterprise

Conchán, 1998

Dear Sirs

EVALUATION COMMITTEE

Public tender for the selection of the enterprise

CITY

With the greatest respect,

I am writing to make you aware that I am participating in the PUBLIC TENDER FOR THE SELECTION OF AN ENTERPRISE TO MANAGE THE ELECTRICITY SERVICE IN THE DISTRICT OF CONCHÁN. I am committed to obtaining the legal status duly recognized in the public registers, within 10 days of the publication of the league table, if I am awarded the contract.

I recognize that failing to make good this commitment will disqualify me from being awarded the contract and entitle the EVALUATION COMMITTEE to proceed according to the stipulations in the book of rules.

Yours sincerely,

....

Mr: ....
Identity card No.: ....
Representative: ....

5.3 Request for service

REQUEST FOR ELECTRICITY SERVICE
No.:

1: ABOUT THE PROSPECTIVE CLIENT
NAME(S) AND SURNAME(S): ....
ADDRESS OF PREMISES: .... No. ....

2: IS THE PROSPECTIVE CLIENT
THE OWNER   A TENANT   A REPRESENTATIVE

3: SERVICE REQUESTED

REQUEST FOR:		TYPE OF SERVICE:
☐ NEW SERVICE (+METER)	☐ TO MOVE	☐ DEFINITIVE
☐ MOVE TO THREE-PHASE	☐ COMPLEMENTARY	☐ PROVISIONAL
☐ CHANGE OF TARIFF	☐ OTHER	☐ OTHER ....

4: ABOUT THE PREMISES

NAME OF THE OWNER: ....

USE OF THE PREMISES: ....

5: DOCUMENTS ATTACHED

COPY OF POLLING CARD

POWER OF ATTORNEY (FOR A REPRESENTATIVE)

OTHER OBSERVATIONS:..

....

....

ALL THIS INFORMATION GIVEN IN THIS REQUEST IS TRUE. I COMMIT MYSELF TO MEETING THE COSTS ARISING FROM ADDITIONAL WORK IF THEY PROVE TO BE UNTRUE.

DATE: ..../.........../.............

....

SIGNED THE PROSPECTIVE CLIENT

....

RECEIVED BY

Notes

1 Rural electrification

1 Organización LatinoAmericana de Energía (OLADE) is funded by the governments of Latin America and the Caribbean and has its headquarters in Quito, Ecuador.

2 The management model

1 In some cases systems belonging to the state have been handed over to municipalities to manage.
2 Occasionally the government gives up the property and hands it over to the community to own and manage.
3 This approach is very convenient for the utility because it deals with one client instead of many, thereby reducing costs to the utility, with the problems of administration also passed on to the community.

3 The Conchán pilot project

1 A programme funded by the Ministry of Agriculture, which in the 1990s gave loans for micro-hydro schemes. The results were not very promising and the programme was finally closed. In 1999 the government finally decreed that all outstanding debts to PRONAMACHCS were cancelled.
2 Before the management model was applied, the municipality who were managing the project applied a flat rate of sol15, and over 25 per cent of users defaulted on payment.

4 Tariff structure

1 BT-5 is the official electricity tariff established by the National Commission of Tariffs (CTE) for rural areas. The CTE is currently a part of OSINERG, the regulating body of Peru's electricity industry.

5 Code of Conduct for the operation and functions of the electricity service

1 Consejo Superior de Contrataciones y Adquisiciones del Estado (CONSUCODE) is the body that regulates and supervises government contracting processes in Peru.

6 Contract between the owner and the micro-enterprise operating and managing the electricity service

1   The Regimen Unico de Contribuyentes (RUC) is a list of the organizations paying tax to the state in Peru.

References

Barnes, D, Van der Plas, R and Floor, W (1997) *Tackling the Rural Energy Problem in Developing Countries*, World Bank, Washington DC.

Carrasco, A (1989) *La Electricidad en el Peru, Politica Estatal y Electrificacion Rural*, Tecnologia Intermedia, ITDG, Lima.

Mariginac, Y and Schneider, M (2001) *Towards a World Efficiency Link (WEEL)*, paper for Sustainable Development International, Strategies and Technologies for Agenda 21 Implementation, WIESE, Paris.

Ministério de Energía y Minas del Peru (1999) *Atlas de Mineria y Energía*, Ministério de Energía y Minas del Peru, Lima.

O'Keefe, P (1996) *Energy Markets in Developing Countries*, paper presented at the Social and Economic Impact Renewable Energy One-Day International Conference, University of Northumbria/ETC UK Ltd.

OLADE (1991) *La Energía en America Latina y El Caribe: Expansión de los Setenta y Crisis de los Ochenta*, Latin American Organization for Energy Development, Quito.

OLADE (1999) *Revista Energética*, 23 (4).

OLADE and World Bank (1991) *Evolución, Situación y Perspectivas del Sector Eléctrico en los Paises de America Latina y el Caribe, Informe Regional, Volumen 1*, World Bank and OLADE, Washington DC.

Platow, J and Goldsmith, A (2001) 'Investing in power and people – a global action plan', *Renewable Energy World*, 4 (6), pp. 47–59.

Sánchez Campos, T (1998) *La Sostenibilidad de los Proyectos de Electrificación Rural*, paper presented at a workshop on Management Models for Photovoltaic Schemes in Peru, organized by the Ministry of Energy and Mines and the National University of Engineering, Lima.

Sánchez Campos, T (2005) *Key Factors in Successful Implementation of Stand-alone Rural Schemes*, unpublished PhD thesis, Nottingham Trent University, Nottingham.

Vorvate, T and Barnes, D (2000) *Rural Electrification in Thailand: The Lessons from a Successful Programme*, draft report, World Bank, Washington DC.

WEC (2000) *Energy for Tomorrow's World – Acting Now!*, WEC Statement 2000, World Energy Council, London.

World Bank (1996) *Rural Energy and Development, Improving Energy Supplies for Two Billion People*, World Bank, Washington DC.

World Bank (1999) *ESMAP Peru Rural Electrification Activity Completion Report*, World Bank, Washington DC.